ROBOTIC FUTURES
建筑机器人建造

Philip F. Yuan, Achim Menges, Neil Leach
袁 烽,(德)阿希姆·门格斯,(英)尼尔·里奇 等著

支持单位
Supported by

  同济大学建筑设计研究院 (集团) 有限公司
TONGJI ARCHITECTURAL DESIGN (GROUP) CO.,LTD.

DADA 数字建筑 设计专业委员会

本研究得到同济大学建筑设计研究院（集团）有限公司重点项目研发基金, 高密度人居环境生态与节能教育部重点实验室自主与开放课题（课题编号：2015KY08）的支持。
Support from key project research funding of Tongji Architectural Design (Group) Co., Ltd. and Key Laboratory of Ecology and Energy-saving Study of Dense Habitat(Tongji University),Ministry of Education . (Grant No.2015KY08) is acknowledged.

INTRODUCTION
介绍

机器人建造在当今的建筑行业的发展中扮演了重要的角色，特别是随着工业4.0时代的到来，基于工业机器人的数字加工模拟和针对不同建筑材料定制化的工具设计，使工厂在大规模生产中经济地制造非单一的建筑预制构件成为可能。在这一新的时代背景下，数字设计建造工具赋予了建筑师更广阔的发挥想象力的空间，并可以亲身投入到建造的流程中。而建筑师对于材料的应用，也从现在被动地接受形态向主动地生产形态发展。这种材料应用的更新手段是对于传统建筑设计的颠覆，而所有的这一切信息都将贯穿于前期的计算设计到后期的机器人建造项目的全过程中。机器人建造除了在实体建筑的加工和建造方面，还在艺术及设计创新领域得到了一定的发展，通过人机交互技术的开发，智能机器人走入日常生活似乎已不再遥远。

本书旨在通过对该领域最新发展动态的纵览，一窥机器人建造对当代建筑实践和艺术设计创新的影响。就目前来看，对于这一领域的探索还局限在世界各大顶尖建筑院校和拥有双重身份的学院派建筑师上，所以要使这一技术全面融入建筑业中，我们还有很长的一段路要走。未来的建筑业将会是什么样？我们并不能给出确切的答案，但本书的这些撰稿人将为我们带来诱人的一瞥——带领我们窥探即将在建筑业中成为常态的材料研究与建造技术。

本书是"数字化设计前沿"系列丛书中的一本。"数字化设计前沿"系列丛书旨在描绘数字技术对建筑学科的影响。2012年，系列丛书中的最初两本——《建筑数字化建造》与《建筑数字化编程》，在袁烽与尼尔·里奇的主编下，以双语形式在同济大学出版社出版。2013年，作为该系列的第三本书作《探访中国数字建筑设计工作营》出版。

我们衷心感谢本书的撰稿人和同济大学、同济大学出版社所作出的贡献；同时，我们衷心感谢袁佳麟的汇编和排版工作，以及黄舒怡、张立名、肖彤、柴华在编译过程中的大力协助。

袁烽，阿希姆·门格斯，尼尔·里奇

Robotic fabrication is beginning to play a major role in the construction industry. Especially with the coming of Industry 4.0 Era, Digital fabrication involving the use of industrial robots and customized design tools for different materials offers the possibility of producing unique pre-fabricated building components within a reasonable price. Digital design and fabrication tools give architects a wider scope for their imagination, and allow them to be personally involved in the fabrication process. The approach towards materials has also developed so that they become the active generators of form rather than simply the passive receptors of shape making. This new method of material application is an improvement on traditional design. All these factors will play a role throughout the whole process of the project, from the early computational design through to the final robotic fabrication. Robotic fabrication has also been developed in art, design and other creative industries. As man-machine interaction develops, so the intelligent robot will come to play an important role in daily life in the not so distant future.

This volume seeks to offer an overview of the impact of robotic fabrication technologies on contemporary architectural practice and design. So far the exploration of this field has been limited largely to the academic environment. There is still a long way to go before these technologies are fully integrated in the construction industry. What will the future of construction look like? We cannot know for sure, but the contributions to this volume offer us a tantalizing glimpse of the kind of material investigations and fabrication techniques that look set to become commonplace in the construction industry.

This volume is part of a series of publications, DigitalFUTURE, that charts the impact of digital technologies on the discipline of architecture. In 2012 the first two books in this series, *Fabricating the Future* and *Scripting the Future*, were edited by Philip Yuan and Neil Leach, and published by Tongji University Press as bilingual editions. In 2013 a further book, *Digital Workshops in China*, was published in the same series.

The editors would like to thank the authors of the articles in this volume for their contributions, together with Tongji University and Tongji University Press. They are also thankful to Crisie Yuan for helping to compile this volume and to Shuyi Huang, Lim Zhang, Xiao Tong and Chai Hua for their further assistance with translation.

Philip F. Yuan, Achim Menges, Neil Leach

CONTENTS 目录

PREFACE 绪论

10 Confluence of High Performance-based Architecture and Robotic Fabrication Industries
    / Philip F. Yuan
    基于机器人建造的高性能建筑未来
    / 袁烽

17 Coalescences of Machine and Material Computation
    Rethinking Computational Design and Robotic Fabrication through a Morphogenetic Approach
    / Achim Menges
    机器生产与材料计算的融合：从形态生成法重新思考计算设计与机器人建造
    / 阿希姆·门格斯

23 Arts-based Robotic Applications through Dynamic Interfaces
    From Fabrication to Performance in the Creative Industry
    / Johannes Braumann, Sigrid Brell-Cokcan
    通过动态界面实现的机器人艺术应用：创新产业中从建造到表演
    / 约翰尼斯·布朗曼，西格丽德·布瑞尔-考根

METHODOLOGY 方法研究

32 A New Building Culture
    Towards a Radical Confrontation between Data and Physics
    Jan Willmann, Fabio Gramazio, Matthias Kohler / Gramazio Kohler Research, ETH Zurich
    一种新的建筑文化：走向数据和物理间的激进对抗
    简·威尔曼，法比奥·格拉马齐奥，马蒂斯·科勒 / 苏黎世联邦理工学院 "格拉马齐奥与科勒" 研究中心

44 A Creative Platform
    Curime Batliner, Michael Jake Newsum, M.Casey Rehm / Southern California Institute of Architecture (SCI-Arc)
    一个创新的平台
    克瑞姆·巴特莱纳，迈克尔·杰克·纽斯曼，M. 凯西·莱姆 / 美国南加州建筑学院

54 Explorative Design and Fabrication Strategies for Fibrous Morphologies in Architecture
    Moritz Dörstelmann, Marshall Prado, Achim Menges / ICD Institute for Computational Design, University of Stuttgart
    建筑纤维形态学的探索性设计与建造策略
    莫瑞斯·多斯特曼，马歇尔·普拉多，阿希姆·门格斯 / 德国斯图加特大学计算设计学院

66 New Craftsmanship Using Traditional Materials
    Philip F. Yuan, Hyde Meng, Zhang Liming / Digital Design Research Center, CAUP, Tongji University
    传统材料的数字新工艺
    袁烽，孟浩，张立名 / 同济大学建筑与城规学院数字设计研究中心

74 Machines for Rent
    Francois Roche, Camille Lacadee / New-Territories
    出租机器
    弗朗斯瓦·罗氏，卡米尔·拉卡迪 / 新领域事务所

CASE STUDIES 案例分析

88 Robotic Stereotomy
The MIT Sean Collier Memorial
J. Meejin Yoon, Eric Höweler / Höweler + Yoon Architecture
机器人切石法：MIT肖恩 • 科利尔纪念碑的建造
尹美真，埃里克 • 霍威尔 / Höweler + Yoon建筑事务所

96 La Voûte de LeFevre
A Variable-Volume Compression-Only Vault
Brandon Clifford, Wes McGee / Matter Design Studio
LeFevre拱：一个可变体量的仅受压拱
布莱登 • 克利福德，威尔斯 • 麦克盖 / Matter设计工作室

106 Shadow Lines
An Experiment in Simulated and Speculative Manufacturing
Bob Sheil, Thomas Pearce, Grigor Grigorov / The Bartlett School of Architecture, UCL
阴影线：一种模拟和推测性建造的试验
鲍勃 • 谢尔，托马斯 • 皮尔斯，格里格 • 格里戈罗夫 / 伦敦大学学院Bartlett建筑学校

112 Clay Robotics
Kwang Guan Lee, Sun Jiashuang / The Bartlett School of Architecture, UCL
机器人粘土打印
康源，孙佳爽 / 伦敦大学学院Bartlett建筑学校

118 Landesgartenschau Exhibition Hall
Robotically Fabricated Lightweight Timber Shell
Tobias Schwinn, Oliver Krieg, Achim Menges / ICD Institute for Computational Design, University of Stuttgart
Landesgartenschau 展馆：机器人建造的轻型木壳结构
托比斯 • 斯科文，奥利弗 • 克里格，阿希姆 • 门格斯 / 德国斯图加特大学计算设计学院

124 Woven Clay
Experiments in Robotic Clay Deposition
Jared Friedman, Heamin Kim, Olga Mesa / Harvard Graduate School of Design
编织粘土：在机器人粘土沉淀上的实验
贾瑞德 • 弗莱德曼，赫敏 • 金，奥尔加 • 麦莎 / 哈佛大学设计研究生院

132 Reverse Rafter
Structural Performance Simulation Based On Wood Tectonics
Philip F. Yuan, Chai Hua / CAUP, Tongji University
反转檐椽：传统木构的结构性能模拟
袁烽，柴华 / 同济大学建筑与城市规划学院

142 Unfolding Topology
Bandsawn Bands
Ryan Luke Johns, Nicholas Foley / Greyshed
展开的拓扑结构：电锯木条椅的制作
瑞安 • 卢克 • 约翰斯，尼古拉斯 • 弗利 / Greyshed设计事务所

PROJECTS OVERVIEW 项目概述

150 Composite Wing & Brass Swarm
Roland Snooks / Studio Roland Snooks & Kokkugia
复合翼与黄铜群
罗兰德·斯怒克斯 / 罗兰德·斯怒克斯工作室 & Kokkugia事务所

156 Robotic Lattice Smock
Andrew Saunders (RPI) & Gregory Epps (RoboFold)
机器人点格褶裥
安德鲁·桑德斯（美国伦斯勒理工学院）& 格里高利·艾比思（RoboFold公司）

160 Design of Robotic Fabricated High Rises
Gramazio Kohler Research, ETH Zurich & SEC Future Cities Laboratory (FCL)
机器人建造高层建筑设计
苏黎世联邦理工学院"格拉马齐奥与科勒"研究中心 & 未来城市实验室

166 Mesh Mould
Gramazio Kohler Research, ETH Zurich & SEC Future Cities Laboratory (FCL)
网格模型
苏黎世联邦理工学院"格拉马齐奥与科勒"研究中心 & 未来城市实验室

170 Automaton
Designing Intelligence
Robothouse, Southern California Institute of Architecture (SCI-Arc)
自动：设计智能化
美国南加州建筑学院机器人实验室

176 Gesture translation & Light Object
Robothouse, Southern California Institute of Architecture (SCI-Arc)
手势转译与光物体
美国南加州建筑学院机器人实验室

180 Domain
Research on Design and Robotic Manufacturing in the Field of Double Curved Geometries
Sandra Manninger, Matias del Campo / Taubman College, University of Michigan
领域：双曲几何领域设计与机器人制造研究
马宁格，马德朴 / 美国密歇根大学Taubman建筑与城规学院

186 Robotic Building
Hypebody, TUDelft
机器人建筑
荷兰戴尔夫特大学Hypebody研究小组

192 The Exploration on the Teaching of Robotic Fabrication and Design
Yu Lei, Xu Weiguo / School of Architecture, Tsinghua University
在机器人建造设计教学上的尝试
于雷，徐卫国 / 清华大学建筑学院

198　Spatial 6D Biomimetic Printing
　　　　DigitalFUTURE Shanghai Summer Workshop, Tongji University
　　　空间仿生结构6D打印
　　　　同济大学上海"数字未来"暑期工作营

202　Edge Chair
　　　　Digital Design Research Center (DDRC), CAUP, Tongji University
　　　边锋椅
　　　　同济大学建筑与城市规划学院数字设计研究中心

206　Topological Surface
　　　　Digital Design Research Center (DDRC), CAUP, Tongji University
　　　拓扑表皮
　　　　同济大学建筑与城市规划学院数字设计研究中心

BIOGRAPHIES 作者简介

PREFACE
绪论

Philip F. Yuan
袁烽

Achim Menges
阿希姆·门格斯

Johannes Braumann, Sigrid Brell-Cokcan
约翰尼斯·布朗曼，西格丽德·布瑞尔-考根

Confluence of High Performance-based Architecture and Robotic Fabrication Industries

基于机器人建造的高性能建筑未来

Philip F. Yuan / CAUP, Tongji University
袁烽 / 同济大学建筑与城市规划学院

机器人作为一种高精度与高效率的加工与建造工具已经广为人知，而随着机器人越来越深入核心地介入各个行业，其意义已经不能仅仅用"高精度"与"高效率"来概括，机器人以及与之紧密联系的智能技术网络带来的强大开放性正在改变虚拟世界与现实世界之间的联系，人与机器、人与人之间的关系正在被重新定义，机器人开始在社会层面上产生影响。从建筑师的角度来说，机器人的介入提供了一种界面，一种数据与动作、虚拟与现实之间的交互界面，这使得建筑师在对从设计到建造的过程的把握更加游刃有余，设计更加自由化，同时也不会缺失建造的合理性，这让建筑师在整个建筑设计与建造过程中达到一个更加自主的状态。另一方面，借助机器人，设计与建造双方面可以形成一个很积极的交互，这将从本质上影响到未来建筑的设计与建造方法以及价值观——高性能植入成为一体化建筑设计与建造方法的核心，同时建筑性能成为评价建筑的重要因素，最终，这将给未来建筑存在的方式带来革新。

高性能建筑未来：从形式驱动到性能驱动

从建筑形式的生成逻辑上来讲，参数化设计正在被重新思考。建筑的性能化目标正在成为参数化设计的重要内容。在历史上，从文艺复兴古典柱式到现代主义新建筑五点，范式转变往往以宣言方式出现，但需要被社会实践检验[1]。如今，建筑师正在从更理性的角度，尤其是建筑性能化以及建筑生产的角度去思考建筑美学以及建筑形式的意义。如何提升建筑与自然、建筑与人的全新关系，并且通过建造来实现这种目标，正在成为建筑学术研究的重要问题。从参数的性能目标出发，探索城市、空间、组织、

The robot is well known as a highly precise and efficient tool for manufacturing and fabrication. However, with the robot being deeply involved in various industries, it is becoming more than just a highly precise and efficient tool. The powerful opportunities afforded by the robot and the closely-related intelligent technology network are changing the relationship between the virtual world and the real one. Moreover, the relationship between humans, and that between humans and machines is being redefined. The robot is exerting an ever greater influence at a social level. From the perspective of architects, the robot offers an interface between data and action, between the virtual and the real so that architects can not only have more freedom to design but also keep control of fabrication. Architects can be more autonomous throughout the whole process from design to fabrication. In addition, the robot promotes positive interaction between design and fabrication, which will influence design and fabrication approaches in the future, as well as basic values. The incorporation of high performance considerations becomes the essence to integrating architectural design and fabrication, and in turn building performance becomes a key factor in building evaluation. Finally, the robot will revolutionize future construction.

The Future of High Performance Architecture: From Form Driven to Performance Driven

Parametric design is now being reconsidered as the redefinition of the parameters. The information of performance-based design is outlining

结构、建造、环境适应性及人群行为等深层次的城市建筑逻辑问题，探讨深层次的城市、建筑生成以及建造的可能性[2]。我们认为，建筑的元素具有参数化属性，并且可以通过机器人数字建造技术的发展，高性能地实现参数的形式意义。

如今，建筑学正朝着"形式追随性能"方向转变[3]。教条式的形式语言，无论是纯粹的建筑几何，还是逻辑算法生成的空间形态，都在遭受到质疑。纯粹的形式主义规则并不能应对真正的城市与建筑问题。除了会导致形式同化，新教条主义带来的建筑无法真正回应环境以及历史文脉带来的丰富性与层次性。随着性能化模拟和数字化技术的出现，其表现的意义也更加直接地对应环境气候关系以及人的行为关系。参数化技术让"形式追随性能"变得更加具有可操作性，这使得形式所能表现出的内在意义逐渐转移到对建筑本身的性能以及存在的伦理意义层面。无论是结构性能、环境性能还是行为性能，都将成为寻找形式意义的有力的出发点。

高性能建筑虽然基于数字技术，但并非仅仅是一种纯粹的形式追求，相反，高性能建筑所追求的是一种通过精准的操作达到人与人以及人与自然的契合融合。一方面，高性能建筑通过研究基于自然规律的仿生学而存在，用最合理的方式、最小的自然资源消耗来实现高效能的建筑；同时，高性能建筑追求的是以更加贴切的人的行为，创造更加舒适、便捷与人性化的空间。因此，高性能建筑最终要实现的是和谐共处和可持续发展的伦理学目标。

将性能模拟和建模与机器人平台连接

性能模拟与生形建模是实现高性能建筑设计与建造一体化的工作基础。性能数据是高性能建筑设计中的核心操作工具，而性能模拟是获取性能数据的最有效手段。基础的性能逻辑（如结构逻辑）可以促成原型的生成，将原型进行几何逻辑参数化后，通过当前的性能模拟软件对不同参数下的原型进行性能分析，可以建立起参数化原型中的控制参数与性能数据之间的逻辑联系，为进一步性能植入原型与多目标性能优化做好准备。如2013年上海创盟国际"自主建构"展览作品"轻拱"中，拱由130多块砌块单元组成，所有的砌块单元都是基于同一种参数化的几何原型，以适用不同的砌块形态要求[4]。

性能建模是对原型进行进一步深化设计的方式。植入方式为基于原型控制参数与性能数据之间的逻辑联系，使用多目标遗传算法对设计原型进行性能优化。将定量化的评价结果反馈到原型研究过程中，对材料组合性能进行优化，并再次作为参数输入原型设计工具包进行设计更新，重复迭代，逐步优化设计。对建筑原型进行性能植入首先需要对多目标的性能优化进行目标设定。环境性能则考虑最优的日照、通风、导热以及声学等环境效应；而结构性能主要关注原型设计对力的承载与传递，使原型在结构方面实现自支撑等特性。如2012年同济大学参加"太阳能十项全能竞赛"的作品复合生态屋中，表皮单元依据受太阳辐射和通风模

the new scope of the geometrical form-finding process. In history, from the classical orders of the Renaissance to the five points of new architecture in Modernism, paradigm shifts have always been presented as a manifesto that should normally be judged in the light of subsequent social practice [1]. Nowadays, from a more rational perspective – and especially from the perspective of integrating design and fabrication– architects tend to adopt a new ethical approach to this new paradigm shift. The challenge of improving the relationship between buildings and nature, and between buildings and human beings has become a significant focus for architecture. In the architectural form finding process, more factors, such as spatial organization, performance modeling, construction methodologies, environmental sustainability, human behavior and other concerns related to the built environment have been taken into account [2]. We argue that the parametric aspects of architecture could not only be defined geometrically by the performance based purposes, but also actualized effectively by linking them to the process of robotic fabrication.

Nowadays, the argument is that the discipline of architecture has switched to the logic of "form follows performance" [3]. The old dogmatic formal approach, no matter whether it is based purely on architectural geometry or is derived by simple computational logic, has been increasingly called into question. It is very difficult for a formal approach to solve complicated architectural problems in the urban environment. This dogmatic approach could not so easily respond dynamically to complex environmental concerns and take

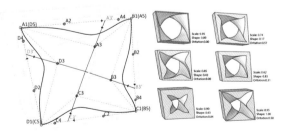

在2013年的"轻拱"项目中砌块单元的参数化原型
Light Vault, 2013: Parametric Prototype of the Component

拟进行优化，实现针对多项环境性能的多目标优化设计表皮菱形构件的加工过程中需要重复地进行大量复杂的竹木板材切割操作以及相对应不锈钢节点定制。考虑到加工的精度和效率要节点定制。考虑到加工的精度和效率要求，大型3轴CNC数控机床被引入加工环节中用于复杂形体的制造[5]。

2014年，在创盟国际完成的合肥"坝上街"项目中，表皮系统的设计由建筑的风压分析所驱动。一方面能够令建筑有独特的美感，另一方面又能够影响建筑环境。表皮由纵向翻转角度不同的铝板阵列组成，基本设计逻辑是：在翻转角度越大的表皮区域，表皮对风流动的阻碍就越大；反之，表皮对风流动的阻碍就越小。这样可以有效控制风在建筑表面的流向以及流速。进一步对建筑表皮进行风压模拟之后，再将风压数据转化为铝板翻转的角度，从而完成立面的设计。在深化设计中，所有的铝板依据安装角度的不同和安装位置的不同进行编号，使得铝板安装过程可以更加便捷。

机器人工具端开发作为实现材料建造的手段

机器人是实现高性能建筑设计与建造一体化中的重要工具，高性能建筑设计中性能优化的结果中呈现的空间复杂性以及表皮渐变特征，对于传统的以手工建造为主的建造技术来说是一个巨大的难题，而机器人建造工具给了这个问题一个正面的解答：机器人平台基于参数的操作模式，使这种高性能建筑的设计与建造真正成为一套连续完整的模式。

基于机器人平台的材料组织是实现性能化运算与数字建造的纽

into account factors, such as the richness of historical context. With the emergence of performance modeling and digital technologies, a window could be opened to set up a relationship with both climate and human behavior. Parametric technology allows the "form follows performance" paradigm to develop programmatically into an ethical concern for architecture. No matter whether it relates to structural performance, environmental performance or behavioral performance computational modeling is set to become a powerful driver in the form finding process.

Although high performance-based design is based on digital technologies, it does not focus purely on form. On the contrary, high performance design pursues the balance and coexistence between humans and nature through a series of precise operations. On the one hand, high performance design tends to learn from biology and actualize effective buildings through the most reasonable methods and with the minimum consumption of natural resources. On the other hand, high performance design also tries to provide more comfortable, convenient and user-friendly spaces. Therefore, the challenge is to achieve such ethical goals, as harmony between architecture and nature, and between architecture and human beings, as well as to promote sustainable development.

Link the Performance Simulation and Modelling to the Robotic Platform

Performance simulation and modeling offer a fundamental methodology in realizing the integration of high-performance design

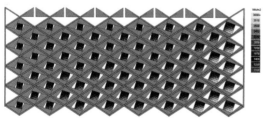

2012年同济大学参加"太阳能十项全能竞赛"的作品——"复合生态屋"，在这项目中对表皮单元进行了受太阳辐射情况模拟与优化。该项目的表皮菱形单元均采用CNC切割加工。
Para-Eco House, Solar Decathlon Competition 2012: The components of the skin are optimized by the acception of solar radiation and fabricated by CNC machine.

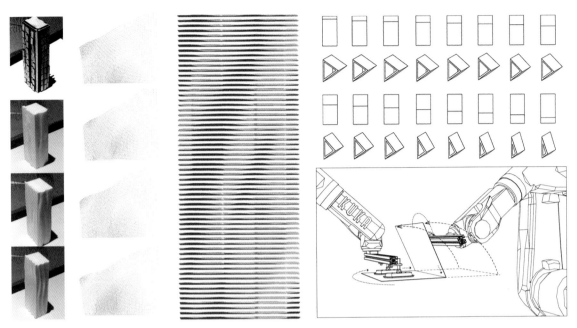

在2014年合肥"坝上街"项目中,立面的风压情况模拟结果生形,该立面铝板使用机器人定制加工。
Bashangjie, 2014: the form of facade is genenrated based on the simulation of the wind pressure and the panels are fabricated by robot.

带。深入量化材料性能,并且通过模拟和参数收集来建立材料基本性能数据库,成为从设计过渡到建造的重要依据。材料性能的研究首先是指收集研究材料的物理和力学固有特性,这些特性既包括材料的强度、刚度、比重、热塑性、热惰性等物理指标,也包括色彩、肌理、质感、气味等感官属性;其次是指对材料加工工艺的研究,材料性能决定了材料的加工方式、材料的几何属性、材料与材料的连接方式。因此,进行材料性能进行研究的同时,对材料加工工艺的研究也不可缺少:包括针对制作工艺、加工手段、连接构造方式等在内的传统工艺的学习和分析,针对不同材料建立材料工艺工具包,作为设计与建造的重要基础。如2014同济大学"上海数字未来"暑期夏令营的"基于材料性能的自主建构"项目中,设计团队选择PLA作为材料,针对PLA材料作为一种非晶体材料熔点不固定的特性,将PLA材料融化后挤出,利用重力在其缓慢凝固过程中形成悬链结构形体,完成具有结构性能的塑形。

就机器人建造工具来说,工具可以分为软件与硬件两个层面。从软件层面,机器人工具所需要的是将建造原型的结合参数转化为机器人动作参数,这种转化的核心在于逻辑的建立与转化,建立几何信息与机器人动作之间的关系。即针对不同类型的几何形体,如平面、单曲面、直纹曲面、双曲面等,将其几何参数,如坐标、曲率、法向量等,依据被加工材料的特性和机器人动作逻辑与条件,转译为相应的机器人加工动作参数,如位置、姿势、速度等。从硬件层面,机器人工具是指针对不同材料和不同加工方式的机器人工具端。机器人能胜任各种加工工作的基础在于其工具端的开放性,使用者可以根据自己的要求更换机器人的工具端,来完成不同的加工任务。工具端

within the fabrication process. Performance data is the key to high-performance design approaches, while performance modelling is the most effective method to obtain performance data. Basic rules (such as structural logics) can generate various prototypes. After the geometrical form finding of the prototype, performance data is introduced under different parameters using the latest performance modelling software, and the logical relationship between the control parameter and performance data in the parametric prototype is established to allow for future performance modeling and multi-objective performance optimization. In the "Autonomous Tectonics" exhibition, the "Light Vault" designed by Archi-Union Architects in 2013 consisted of over 130 block units, all of which were based on one parametrically controlled geometry prototype to meet different requirements for block forms [4].

Performance modeling is the next step for the further design development of the prototype. It seeks to optimize the performance of the design prototype through multi-objective genetic algorithms based on the relationship between the control parameters and performance data of the prototype. The quantitative evaluation result will give feedback to the prototype study, which will optimize the composite performance of the materials, and then input them again into the prototype design toolkit for design renewal and iteration. For performance modeling to building prototype, the first step is to set up goals for multi-objective performance optimization. Environmental performance studies needs to be made of optimal sunlighting conditions, ventilation, thermal conductivity, and acoustic environment effects, while structural performance analysis needs to

开放让建筑师能够更本质更直接的介入到建造过程中来：例如2014同济大学"上海数字未来"暑期夏令营的Struc[Punch]ure项目中，设计团队为了依据结构受力线在金属板上加工出线型纹理，设计了一款用于机器人上的金属锤击加工工具，通过快速锤击动作在金属板上连续打点，形成线型纹理。

就机器人建造方法来说，机器人建造提供的是一个开放的、数字化的工作平台，这种开放性是机器人建造方法的核心。建筑师对机器人建造的理解不能仅仅的认为机器人是一个可以代替手工加工的高精度机器，其基于数据的高度开放性与可适应性才是建筑师应该认识到的一点。在这个平台上，所有的工具都可以被选择，依据加工步骤，加工工具随时更换；所有的加工指令都依据几何逻辑与建造逻辑被数字化，可以通过修改参数进行调整。

基于机器人建造的建筑产业化思维

随着机器人建造引领的数字化设计与建造技术发展，先进的设计管理理念，先进的设计支持工具，先进的加工设备，不断出现的新材料，大大促进了建筑向模块化、预制化、可定制化、产业化方向转变。在基于设计与建造一体化的建筑产业化背景下，建筑业由劳动力密集型行业向设备密集型行业转变，由粗放式生产向更精确化控制加工方式转变，由大量现场作业向大量工厂定制化转变。

如上海城建集团完成的上海市浦江基地05-02地块保障房工程项目，在深化设计阶段，BIM平台完成了PC 零件库、碰撞检查、动态施工仿真、深化设计自动化。在该阶段通过Revit 模型导入到Tekla里，为深化设计阶段预留、预埋打好基础，并进行钢筋深化的设计。首先，由于构件是在工厂事先生产好而后运输到施工现场安装，从而对深化设计提出了相当高的要求，利用Revit 的模型导入到Tekla 里，把每个节点都进行智能化的碰撞检查后，避免了设计、构件制作以及现场施工矛盾。其次，上海市地下空间设计研究总院有限公司经过对Tekla软件的研究后，自行进行了软件的二次开发，完成了自动配筋过程，大大地提高了工作效率，并且利用基于BIM的Autodesk Revit系列软件在建筑设计阶段自动出图，节约了人力成本并缩短了作图周期。在构件生产中，BIM完成了模具设计自动化、生产计划管理、构件质量控制，对改进传统构件生产模式有很好的补充作用。通过RFID 芯片将虚拟的BIM 模型与现实中的构件联系起来，实现了构件生产的集约型管理，同时也使构件有了属于自己的身份证，为把控构件的全寿命周期提供了重要条件 [6]。

设计方法与工具革新下的机器人建筑未来

这种革新对建筑师来说，首先意味着建筑师对原型设计与建造的介入进入到了一个更深的层面，即原型生产层面。这里指的原型生产并非单纯的在原型设计完成后进行成产，而是将材料性能在设计之初就纳入考虑的范围内，并贯穿始终。在设计完

address questions such as load bearing and load transmission in the prototype design in order to endow the prototype with the capacity to be self-supported. For the Para-Eco house designed by the CAUP of Tongji University for the 2012 Solar Decathlon, the skin was designed through solar radiation and ventilation modelling to realize optimal multi-objective design in terms of multiple environmental performances. Lots of complicated bamboo plate cutting and customization of corresponding stainless steel nodes were required during the fabrication of the rhomboid components of the skin. Tailor-made nodes were required, and a large 3-axes CNC machine was employed to create the complex prototypes so as to obtain the required manufacturing accuracy and efficiency [5].

In the Hefei Bashangjie Project designed by Archi-Union Architects in 2014, the facade system was designed based on the analysis of wind pressure on the building. The facade system not only endows the building with its unique appearance, but also affects building's environmental performance. The facade is composed of arrays of aluminum plates with different vertical flip angles. The logic for this design is as follows: the greater the flip angle in the skin area, the greater the wind resistance, and vice versa. Thus, the direction and speed of the wind on the building's surface can be effectively controlled. After further wind pressure analysis of the skin, the wind pressure data is converted to control the tilt angle of the aluminum plates, which are incorporated into façade design. In the design development, all the aluminum plates are numbered according to the different installation angles and locations to ensure a more convenient and rapid installation process.

Robotic Tool Development as An Approach to Material Fabrication

The robot plays a significant role in the integration of high-performance design and fabrication. The optimization of high performance buildings tends to be more complex space and differential surface characters. It poses a great challenge to traditional labor-based construction technologies, while robotic fabrication could solve this problem. Actually, the robotic platform is based on a parametric driven system, and it makes sense to integrate design and fabrication processes based on this new design methodology.

Material organization based on the robotic platform links performance-based computation to digital fabrication. Through quantification of material performance and the establishment of a database for different basic material characteristics, we could set up the key basis for transition from design to fabrication. The study of material performance first entails to research into the internal physical and mechanic features of the materials, including material strength, stiffness, specific gravity, thermodynamic plasticity, thermal inertia and other physical properties, such as

成之后，借助机器人建造平台，建筑师可以直接参与到原型生成的过程中。这将使建筑师对整个从设计到建造的过程的掌控度达到的更有挑战性的状态。

其次，设计公司的核心竞争力将不再只是建筑设计与建筑工程方面，基于新的机器人建造平台的高性能建筑材料研究与机器人加工工具研发将成为新的核心竞争力。尤其是对于研究型的中小型建筑事务所来说，强大的新材料和新工具的研发能力将在高性能建筑市场中成为独特而有力的竞争手段。

最后，在工业4.0时代背景下，网络化信息传递以及性能化定制成为了数字化工业生产的第一要义，因此基于机器人平台和网络平台的建筑数字化设计与建造将成为无可避免的趋势。云端设计与建造资源整合、高度定制化的设计与预制的需求与服务以及高度信息化的设计与建造手段将会是机器人高性能建筑的未来。

松江名企艺术产业园区（2014）：预制与定制结合进行的建筑数字建造
Songjiang Art Campus (2014): Digital construction adapting pre-fabrication and customization

color, texture, smell and other sensory features. Secondly, it entails research into material manufacturing techniques. As material performance determines the manufacturing method, geometric attributes and connection methods, it is essential to research material manufacturing techniques during the material performance study and to analyze the fabrication process, manufacturing and connection methods, and other aspects of traditional craftsmanship, and to establish a craftsmanship toolkit for different materials as the key basis for the design-fabrication process. In the 2014 DigitalFUTURE Shanghai summer workshops, the project – "Autonomous Tectonics", designed by students of Tongji University, PLA was chosen as the material. As an amorphous material without any fixed melting point, PLA was first melted and then extruded to form a catenary structure with the aid of gravity as it gradually solidified to take its final complete structural form.

With robotic fabrication tools, both software and hardware are involved. On the level of software, robotic tools need to translate binding data of building prototypes into robotic movement data; the core of this translation lies in the establishment and conversion of logics and the establishment of a relationship between geometric information and robotic movements. That is to say, geometric parameters like the coordinates, curvature and vectors of different geometries such as planes, single curved surfaces, ruled surfaces and hyperboloids are translated into corresponding robotic fabrication movement data (such as position, motion and speed) based on the properties of manufacturing materials and the logic and conditions of robotic movements. On the level of hardware, robotic tools could be subdivided into various categories, depending on the different materials and fabrication requirements. The robots' capacity to perform various fabrication tasks depends on its various end tools. As an open system, the robotic toolhead enables architects to be involved into the fabrication process in a more direct and essential manner. For instance, in the Struc[Punch]ure project completed by students from Tongji University as part of the 2014 DigitalFUTURE summer workshop, the team designed a metal hammering tool fixed to robot arms to produce a linear texture on metal panels following structural stress lines; this tool was able to generate linear textures by rapidly hammering continuous dots on the metal panel.

In terms of robotic fabrication methodologies, what robotic fabrication provides is a digital open platform; it is this openness that is most significant in the robotic fabrication process. Architects should understand that robotic fabrication is more than just a high precision version of a manual fabrication machine; instead they should be aware of the high levels of data-based openness and adaptability of robotic fabrication. On this platform, all the tools can be selected or replaced based on the specific manufacturing steps; all the processing instructions are digitalized based on geometric and fabrication logics, and can be adjusted by revising constraints.

Building Industrialization Based on Robotic Fabrication

With the development of digital design and fabrication technologies using robotic fabrication, advanced design management concepts, supporting design tools, processing machinery, and emerging new materials are all factors that significantly allow the architectural profession to transform to modular, prefabricated, customized and industrialized construction. In the context of industrialized construction that integrates design and fabrication, the construction industry is at the point of shifting from labor intensive to industrially intensive strategies, from ready-made production to customized production, and from large scale in situ fabrication to large scale factory prefabrication.

We can take the example of the affordable Housing Project in Plot 05-02, Pujiang Base, Shanghai completed by Shanghai Urban Construction Group. PC components, collision checks, dynamic construction simulation, and the automatic design development were all completed on a BIM platform at the design development stage. At this stage, information was imported into Tekla from a Revit model, laying a solid foundation for space provisions, embedding and reinforcement. First of all, the components were prefabricated in the factory and then transported to the construction site for installation; this required more precise design development. So the information was imported into Telka by using Revit models to perform intelligent collision checks for each and every node, in order to avoid conflicts between design, component fabrication and site construction. Secondly, the Shanghai Underground Space Design & Research Institute carried out secondary software development after studying the Tekla application, and completed an automatic reinforcement process, which significantly improved construction efficiency. The BIM-based Autodesk Revit software program was utilized for generating computerized drawings, thereby saving labor costs while reducing drawing preparation time. During the component fabrication, the automation of the BIM design model, production plan management and component quality control tasks all played a supporting role in improving traditional methods of component production, achieving efficient management while providing each component with an identity, and importantly creating the conditions for the full lifecycle control of components [6].

Robotic Futures: The Introduction of New Design Methodologies and Fabrication Tools

For architects, the introduction of new design methodologies and fabrication tools means first of all that their involvement in prototype design and fabrication reaches a new level in terms of the prototype production. Here prototype production means more than just the fabrication following completion of the design phase. Rather, it means taking into account material performance at the very beginning of the design process and throughout the entire fabrication process. After completing the design, architects can directly engage in the process of prototype production with the help of the robotic fabrication platform. This will increase architects' control over the entire process from design to fabrication at a more advanced level.

Secondly, the key contribution of design firms will no longer be confined to architectural design and engineering aspects. Instead, research into high performance building materials based on new robotic fabrication methods and the research and development of robotic fabrication tools will be their new contribution. In particular, for small and medium sized research-based architectural firms, research and development of new materials and tools will become one of their main contributions in the high performance building market.

Lastly, in the context of Industry 4.0, networked information transmission and performance-based customization become the top priorities for digital industrial production. As a result, the integration of digital design and fabrication based on robotic platforms is brightening up the future. The confluence of cloud computing, resource aggregation, customization, prefabrication, informational design and fabrication approaches in the architectural profession will lead to a high-performance based robotic future.

参考文献 / References：

[1] 袁烽. 数字化建造——新方法论驱动下的范式转化 [J] . 时代建筑，2012(2)：74-79.

[2] Patrik Schumacher, "Parametricism: A New Global Style for Architecture and Urban Design" in Neil Leach (ed.), Digital Cities, Architectural Design ,Vol. 79, No. 4, 2009(7/8): 14-23.

[3] Bruce Kapferer, Angela Hobart. Aestheticsin Performance: The Aesthetics of SymbolicConstruction and Experience. Berghahn.Books, 2006.

[4] Philip F. Yuan, Hao Meng and Pradeep Devadass, "Performative Tectonics.", Robotic Fabrication in Architecture, Art and Design. Springer International Publishing, 2014.

[5] 袁烽，钱烈. 基于环境性能的适应性建筑设计——以2012年欧洲太阳能竞赛参赛作品复合生态屋为例 [J] . 住区，2013(6)：71-76.

[6] 上海城建集团，上海市地下空间设计研究总院有限公司. BIM在预制装配式住宅项目中的应用 [J] . 建筑，2013(11)：45-47.

Coalescences of Machine and Material Computation
Rethinking Computational Design and Robotic Fabrication through a Morphogenetic Approach

机器生产与材料计算的融合
从形态生成法重新思考计算设计与机器人建造

Achim Menges / ICD Institute for Computational Design, University of Stuttgart
阿希姆·门格斯 / 德国斯图加特大学计算设计学院

"计算",最基本的含义是对信息的处理[1]。在这种意义上,二进制数字领域内操作的机器过程,以及复杂物理领域内操作的材料过程,都属于计算过程。当代设计对前者尤为青睐,对后者在建筑领域中也有零星的研究。然而,对于机器计算和材料计算交迭的可能性,仍旧是知之甚少的是一个未知领域。在交迭领域内,两者在设计过程中不仅仅是简单的共存关系,而是密集的交互关系。两者的交互使得在设计计算中整合材料信息(即源于材料性能的特质和源于材料物质性的限制)成为一种必须。它提出了一个概念,材料不应像当前计算设计方法中那样,被动地接受形态,而应主动地生成形态,使我们能够在形态生成设计范式中探索新的性能及建筑可能性[2]。

信息、生形和物质化在生物界中存在着本质的联系。三者间复杂的相互作用产生了生物化的过程。生物体潜在信息之间的关系——基因型,以及同特定环境互动而逐渐形成的形态——表现型,都已经得到了广泛研究。最初,进化生物学认为,基因序列是自然系统从无限可能性中选择形态的首要决定因素;但随着对基因知识的积累,我们开始认识到基因序列并不是形成生物形态的唯一驱动因素。依据最近的科学研究发现,基因组是生物工程决定论的看法迅速被弱化,材料操作的重要性逐步被认可[3]。更加显而易见的是,进化过程是在材料约束下进行的,这意味着形态生成的"调色板"受基因和物质性的双重影响。此外,大自然充分利用了材料的固有属性:生物系统广泛借助当地材料进行交互作用,这种按自然法则计算生成的形态,引发了自组织结构及涌现形态的产生。生物系统中大量存在已知的无生命体的图案成型现象强有力地证明了这一点[4]。

Computation, in its most basic meaning, refers to the processing of information [1]. In this way, both machinic processes operating in the binary realm of the digital, as well as material processes operating in the complex domain of the physical, can be considered computational. While there is a strong bias towards the former in contemporary design, sporadic investigations of the later have also occurred in architecture. What has rarely been explored, though, is the largely unchartered territory where machine computation and material computation potentially overlap, where they don't simply co-exist, but intensely interact in the design process. Operating at this overlap entails integrating material information – that is, behaviors originating in materiality and constraints emanating from materialization – in design computation. It suggests a conception of material not as a passive receptor of shape, as emblematic for current approaches to computational design, but as an active generator of form, which enables the exploration of novel, performative capacities and architectural possibilities within a morphogenetic design paradigm [2].

Information, formation, and materialization are inherently related in living nature. Through their complex reciprocities, the processes of biological-becoming arise. The relation between an organism's underlying information set, the genotype, and its manifest form unfolding from its interaction with specific environment, the phenotype, has been extensively studied. Initially, evolutionary biology seemed to propose that the genetic code is the primary determinant for what natural systems are actualized from an

在本质上，基因序列和材料构造组成了整体关系。

同自然系统的交互特征相反，在建筑领域，信息、生形和物质化之间通常呈线性关系。至少，就形态生成与物质化而言，该过程是单向且层次化的。尽管材料在生物形态的生成过程中扮演着主动角色，但是在建筑领域，材料通常被视为形态的被动接受者，接受由其他因素已然确定的形态。迄今为止，在建筑领域普遍应用的机器计算也未曾改变这一状况。尽管，计算设计的整合特征已被广泛利用以便纳入功能、结构、环境或经济信息，但是材料却几乎从未被视作形态生成的驱动因素，更别提对材料加以应用。看来，长久处于主导地位的以形态为导向的设计表达与生形技术似乎仍旧是当代设计思维的先入为主的方法，这种设计技术基于明确的几何定义形式，在大多数CAD软件中运用这种方法定义与扩展形式。即使在其他先进的、以

infinitely vast space of possibilities. However, as knowledge about genomic information accumulated, we began to recognize that it is not the only driver in the genesis of biological form. With the understanding of the genome as the definite blueprint for biotic construction rapidly eroding in light of recent scientific findings, the critical importance of material processes is becoming recognized [3]. It is ever more apparent that evolution operates within material constraints, meaning that the palette of possible formations is both genetically and physically defined. Moreover, nature substantially capitalizes on material innate capacities: biological systems extensively utilize local material interactions, which is a form of physical computation giving rise to self-organizing structures and emergent forms. The large numbers of known, pattern-forming phenomena of non-living nature that occur in biological systems provide strong evidence for this [4]. In nature, instructive code and material construction constitute an integral relation.

In contrast to the reciprocities characteristic of natural systems, in architecture, the relation between information, formation, and materialization is typically linear and, at least with regard to the genesis of form and its materialization, one-directional and hierarchical. Whereas material plays an active role in the generation of biological form, in architecture it is most commonly conceived as a passive receptor of otherwise determined shape. Thus far, the advent of widespread and increasingly ubiquitous use of machine computation in architecture had seemingly very little effect on this condition. While the integrative character of computational design has been extensively utilized for the inclusion of programmatic, structural, environmental, or economic information, material information is hardly ever considered, let alone employed, as a generative driver. It seems as if the age-long predominance of shape-oriented representational design techniques based on explicit geometry and their direct, conceptual extension in most contemporary CAD packages still preconditions contemporary design thinking. Even in otherwise progressive and behavior-oriented design approaches, materiality is still conceived as a passive property of shape and materialization understood as being subordinate to the creation of form.

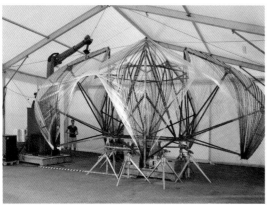

2012 ICD/ITKE研究展馆：该项目通过机器人复合材料加工的综合方法探索建筑与结构的可能性。结构性的黑色碳纤维作用在透明玻璃纤维表面，在力的驱动下，生成了展馆独特的外壳。空心机器人纤维缠绕流程利用纤维计算成形的材料特性，克服了通常建造中对精密模具的需求。建造中，纤维材料在彼此间的相互作用下，生成了展馆的双曲面表皮。
ICD/ITKE Research Pavilion 2012: The project explores the architectural and structural potentials offered by an integrative approach to robotic composite fabrication process. The pavilion's distinctive shell articulation originates from the force-driven pattern of the structural, black carbon fibres on the transparent glass fibre surfaces. The coreless robotic filament winding process overcomes the usual requirement for elaborate moulds by utilizing the fibres' capacity to materially compute shape, as the pavilion's double curved surfaces emerge from the interaction between the fibres during fabrication.

Another reason for the lack of materially-informed design computation may be the difficulty of developing appropriate design methods capable of navigating the narrow path between under-determining material specificity, which leads to a lack of rigor and, consequently, operativeness, and over-constraining material properties and boundary conditions, resulting in both the premature convergence of, and the lack of, exploratory potential. For the identification of an operative and explorative methodological spectrum, two precursors of employing material computation in design may be relevant. One the one hand, Josef Albers' material studies for his Vorkurs at the Bauhaus (Dessau) and, later, at the Black Mountain College (North Carolina) established a precedent

行为为导向的设计方法中，材料的性能仍然被视为形态的一种被动属性，而材料的物质性则被认为从属于形态生成。

缺乏基于材料的设计计算的另一原因是：在未经证明的材料特性（由此导致精确性和可操作性的缺乏）与过度约束的材料属性、边界条件之间存在空隙。我们很难发展出合适的设计方法去跨越这一空隙，因此导致探索潜能的缺乏或不成熟聚合。在验证这种有效方法论方面，有两位相关的将材料计算运用于设计中的先驱者。其一，约瑟夫·阿尔伯斯在包豪斯（德绍）及之后在黑山学院（北卡罗莱纳州）的材料研究基础课程中开创了以材料实验丰富设计流程的先例。阿尔伯斯反对基于专业技术知识建立的物质化过程，称其扼杀了创造性。与此相反，他认为材料变化本身为发展新的结构模式和建筑革新提供了一个创造性的契机 [5]。材料研究作为课程的重要部分之一，并不是量化的模型或概念表达，而是材料性能在时间与空间中的暂时展开。在设计师的手中，这些展开带来了未来的无限可能，钻研出迄今为止未曾被挖掘的设计潜能。

其二，建筑大师弗雷·奥托在斯图加特大学研究所进行实验时广泛采用了他称之为"找形"的实验方法，这可被视作基于材料的指导性设计方法的另一实例。奥托广泛调查了各种材料系统（从诸如网壳等精确定义的结构到沙子等自然形成的颗粒状物质）以研究其自生形能力。奥托系统性地研究了不同材料系统的行为特征以找出特定的形态，这些形态是外在作用力和内在约束力在系统内相互作用至平衡状态时的外在表现[6]。这些材料实验不应与后期设计过程中采用的那些极其精确、繁复的实体结构模型相混淆（如慕尼黑奥林匹克体育场屋顶的光度分析索网模型）。与这些采用物理计算作为工程验证的模型相反，奥托初始的研究宗旨是探索从材料性能出发生成建筑设计的可能性，以突破以往追求形式或空间的设计方法。

将机器生产与材料计算之间的交叉视作建筑咨询与技术创新要求的潜在领域，从本质上说，这可以看作是一种跨学科的设计方法。然而，在此背景下，区分建筑学的材料计算与工程学的材料模拟就显得尤为重要。由于设计和工程技术的需求和特点各不相同，基于材料的计算设计并非简单等同于在设计过程初期采用材料模拟技术。工程模拟追求的是可确定性和精确性，因此简化了模型的已知成分和基本属性，并要求所有的边界条件被良好的定义。相反地，在建筑设计决定性的早期阶段，我们需要一种探索性的技术来创造广阔而开放的设计空间，这一技术能够快速进行生成、变异和进化以适应不断发展的设计意图，基于扩展状态下不完整的信息集，向前推动设计的探索与发现。[7]因此，设计中对材料计算的预期和对材料属性的模拟是一种互补而非竞争的方式。

除了在设计计算中植入材料性能外，材料物质性，因其功能可见性和限制因素，也同样可以成为主动的设计驱动力。由此而论，空间设计与生物物理理论中"形态空间"的概念之间可以建立起十分有趣的关系。形态空间作为计算和概念工具，能够在多维参数

for the possible enrichment of design processes through material experimentation. Albers rejected established processes of materialization based on professional craft knowledge, claiming that they stifle invention. Instead, he identified material behavior itself as a creative domain for developing new modes of construction and innovation in architecture [5]. The material studies, which were undertaken as a vital part of his courses, were conceived not as scalar models or representations of cerebral constructs or ideas, but as temporary un-foldings of material behavior in space and time. In the designer's hands, these un-foldings carried multiple possible futures and bore hitherto-unsought design potentialities.

2013-14 ICD/ITKE研究展馆：展馆的双层壳体证明了为此项目研发的机器人空心纤维缠绕流程的建构潜力。在每一个多变的壳体构件形态背后，是仿生原理、材料特性、建造可行性、结构性能和空间特质的整合。为了建造这种复合构件，团队研发出两台机器人协同进行空心纤维缠绕的创新流程。为了响应当地壳体结构特征，36个构件在尺度、几何形式、纤维排布上大有差别，但仍可以凭借同一个机器人装置完成生产。
ICD/ITKE Research Pavilion 2013-14: The double-layered shell of the pavilion demonstrates the tectonic potential of the robotic coreless filament winding process specifically developed for this project. In the variable component morphology of the shell the biomimetic principles, material properties, fabrication affordances, structural behaviour and spatial characteristics are synthesised. For the production of the composite elements a novel coreless filament winding process for a dual robot set-up was developed. In response to local shell characteristics the 36 components differ considerably in size, geometry and fibre layout, but can all be produced with the same set-up.

空间内描述和关联起生物体表现型之间的巨大差异 [8]。它们构成了由多维坐标轴（每一坐标轴与形态的某一可变参数相关）定义而成的形式空间，同时成为进化生物学话语中的著名概念——自适应景观的基础。此处特别值得关注的是，经验主义形态空间（由所有已知变量的映射定义而成的追溯性空间）和理论主义形态空间（由所有可变性定义而成的前瞻性空间）之间的区别 [9]。计算设计空间成为理论形态空间中所有可能形态的一种子集，这一概念开启了描绘可能的几何形态空间及给定环境中可成形空间之间关系的可能。这些可成形的空间，即所谓的"机器形态空间" [10]，之后便可通过生成性计算设计技术进行操作与探索 [11]。

这种"机器形态空间"设计方法不仅能够描述特定的机器人建造装置，也能够用于研究机器人、感应器以及工具头的改变是如何导致设计空间的变化，缩小或扩张的。基于当前如此系统而复杂的"可建造"定义，我们亟需发展出一种探索空间性能化的设计方法。在此情境下，"仿生学设计方法"显得特别有潜力——因为大自然将持续分化"建筑元素"作为实现高效而性能化材料系统的关键策略之一。而在此类生物系统中，形式、材料与性能之间的潜在关系，对建筑师和工程师来说通常是违反直觉的。因此，自下而上的仿生学设计流程需要跨学科的设计协同，需同生物学家以及相关自然科学家进行密切合作，由此我们才能研究、建立并最终抽象出自然结构和系统的生成法则。通过筛选仿生学的性能化法则，探索出生形差异化范围内可能的形态空间。

基于代理元的建模方法是研究机器形态空间的计算设计的一种极具潜力的方法。在此，"代理元"代表了三维设计空间中有着明确几何形体的实体，同时代表了N维度形态空间中的某一位置。在设计中，若某一结构或系统的一部分定义了一个多智能体系统，而其中每一智能体组成一种建筑元素的话，则这种相互关系可以被工具化。当智能体系统的行为引擎驱动这些建筑元素去探索三维设计空间时，它们在机器形态空间中相应的位置将被登记，以确保它们在可成形区域范围内活动。鉴于这一智能系统包含了可扩展的方法论体系，除之前提到的建造、装配和施工因素之外，其他设计驱动力，如结构、经济或生态标准等都可被整合进设计之中。与当前建立的基于知识的设计体系相反，这种基于行为生成建筑形态的方式是真正探究式的，它使我们能够对机器人建造进行以询问为导向，以实验为基础的研究，探寻拓展设计空间的机器人建造如何为建筑提供创新的空间、结构和生态潜力。

除了在数字领域内的设计探索外，由于基于代理元的建模方法能够整合设计和机器控制领域，因此，基于代理元的系统也提供了直接连接到物质世界的可能性。随着机器人不再依赖于一系列置于机器控制代码中确定的建造指令，而是愈发能够实时对它们所处的环境进行感知、处理和响应，从工业化建造到行为导向建造的转移正逐渐显现。这需要形态生成与物质化具有全新的聚合点，于此，设计不再完全存在于数字领域之中，而

On the other hand, Frei Otto's extensive series of experiments at his institute at the University of Stuttgart, employing what he called form-finding methods, may serve as an example of the other end of the spectrum, of an instrumental material-informed design approach. Otto investigated a vast number of different material systems, ranging from precisely defined structures such as grid shells, to naturally found granular substances such as sand, in order to study their self-forming capacities. He systematically studied their behavioral characteristics to find specific forms, which were manifested as the equilibrium state between external forces and internal restraints acting upon the system [6]. These experiments should not, as is often the case, be confused with the extremely precise and elaborate physical, structural models that followed later in the planning process, like the photometrically analyzed cable nets for the roof of the Munich Olympic stadium. In contrast to these models, which employed physical computation as a means of engineering verification, Otto's initial studies were exploratory inquiries into the possible points of departure for developing architectural designs through material behavior, rather than through the determination of form and space.

Considering the overlap between machine and material computation as a potential domain of both architectural enquiry and technological innovation requires, by nature, an interdisciplinary approach to design. However, in this context, it is critical to distinguish between material computation in architecture and material simulation in engineering. A call for materially informed computational design cannot simply be equated to a call for employing material simulation earlier in the design process, as the demands on, and characteristics of, design and engineering techniques are too different. Engineering simulation seeks determinability, precision, and, as a consequence, reduction to the known constituents and fundamental properties of a model, which requires all boundary conditions to be well defined. In contrast, in the decisive, early phases of architectural design, it is necessary to navigate a vast, open-ended design space, which requires explorative techniques capable of rapid generation, variation, evolution, and adaptation to continuously developing design intent – an advancing oscillation between searching and finding based on an expanding, yet naturally incomplete, set of information [7]. Thus, the anticipation of material computation in design and the simulation of material properties are not competing but complementary methods.

In addition to embedding the characteristics of materiality in design computation, the affordances and constraints of materialization can also be an active design driver. In this context, an interesting relationship between the notion of design space and the concept of "morphospace" in theoretical biology can be established. Morphospaces serve as computational and conceptual tools that allow for describing, as well as relating, the vast variance of

是通过计算建造的探索性过程展露于物质世界。因此，在日益复杂的电脑-实体建造系统中，建筑学的机器未来将成为一种综合性的设计与建造，一种真正的形态生成模式——探索构成我们建筑环境的材料世界。

organismal phenotypes in a n-dimensional parameter space[8]. They constitute the formal spaces defined by multidimensional axes, each of which corresponds to a variable parameter of morphology, and also underlie the prominent notion of adaptive landscapes within the discourse of evolutionary biology. Of particular interest here is the distinction between the empirical morphospace, a retroactive space defined by the mapping of all known variation, and the theoretical morphospace, a prospective space of all possible variability[9]. The conception of a computational design space as a subset of the theoretical morphospace of all possible forms opens up the possibility to delineate between the space of geometrically possible forms and those that are producible within a given fabrication environment. This space of the fabricable, the so called machinic morphospace[10], can then be navigated and explored through generative computational design techniques[11].

The machinic morphospace method allows not only for describing the design space of one specific robot fabrication set-up, but also for the investigation as to how changes to the robots, effectors or tools leads to a shift, contraction or expansion of this design space. With such a systematic and comprehensive definition of the fabricable at hands, a suitable method for exploring the performative regions of this design space needs to be developed. Biomimetic design methods seem to be particularly promising in this context, as nature employs the continuous differentiation of its "building elements" as one key strategy for achieving highly resource-effective and performative material systems. As the underlying relation between form, material and performance of such biological systems are often counterintuitive for architects and engineers alike, bottom-up biomimetic design processes require working in multidisciplinary design collaborations, in this case with biologists and related natural scientists. Thus, the principles of natural structures and systems can be investigated, established and ultimately abstracted. In this way, the morphospace of the possible range of fabricable differentiation can be explored through the filter of the performative principle of biomimetics.

2014 Landesgartenschau展馆：该项目是"机器人木构"研究的成果，是世界上第一座由桦木胶合板构成的机器人分段建造的木结构壳体。壳体形态源自对机器形态空间的基于代理的探索。243片几何各异的桦木胶合板以及它们相互间的连接节点都是由一台7轴机器人加工而成的。设计和建造的整合确保了项目的高精度，所有板材之间的平均偏差只有0.4mm。
Landesgartenschau Exhibition Hall, 2014: The demonstrator building of the research project "Robotics in Timber Manufacturing" is the world's first robotically fabricated segmented timber shell with a primary structure of beech-plywood plates. The shell morphology is derived through an agent-based exploration of the machinic morphospace. The timber shell's 243 geometrically unique beech plywood plates and their intricate joints were produced on a 7 axis robotic fabrication set-up. The integration of design and fabrication resulted in high manufacturing accuracy, with a mean deviation of only 0.4mm across all measured plates.

One particularly promising approach for the computational design exploration of the machinic morphospace is agent based modeling. Here agents can represent both an entity in the 3D design space of explicit geometry and simultaneously a location in the n-dimensional morphospace. In design, this interrelation can be instrumentalized if the parts of a structure or system define a multi-agent system, where each agent constitutes one building element. While the behavioral engine of the agent system drives the building elements to explore the 3D design space, their corresponding location in the machinic morphospace is registered and affects their behavior to stay within the domain of fabricability. As such agent systems comprise an extendable methodological framework other design drivers such as structural, economic or ecological criteria can be integrated in addition to the aforementioned aspects relating to fabrication, assembly or construction. As opposed to established knowledge-based design systems, this behavioral approach to the genesis of architectural form can be truly explorative. Thus it enables an enquiry-oriented and experiment-based investigation of how the vastly extended design space of robotic fabrication opens up novel spatial, structural and ecological potentials for architecture.

Beyond the design exploration in the digital realm, agent based systems also offer the possibility to connect directly to the physical world, because agent based modeling has the capacity to merge the domains of design and machine control. As robots no longer remain dependent on a clear set of manufacturing instructions cast in determinate machine control code, but instead are increasingly capable of sensing, processing and responding to their "environment" in real-time, a shift from instructional to behavior-based fabrication is looming on the horizon. This may entail a novel point of convergence of form generation and materialization, where design is no longer fully anticipated in the digital realm, but instead unfolds in the physical world through explorative processes of computational construction. Thus, in the context of ever more sophisticated cyber-physical fabrication systems, the robotic future in architecture will be one of synthesized design and making, a truly morphogenetic mode of exploring the material world that constitutes our built environment.

鸣谢 / Acknowledgements:

本文写作的基础是作者在2012年发表的一篇论文的扩写。要感谢所有斯图加特大学计算设计学院参与该领域研究的师生、科研人员和校内外的合作者。还要感谢"仿生竞争网络"的合伙人以及所有支持并赞助过该领域研究的赞助商。

The text comprises an extended version of a position paper first published in 2012. The author would like to thank everyone who contributed to further developing this line of research, especially the researchers at the Institute for Computational Design, the collaborators of the Institute of Building Structures and Structural Design at the University of Stuttgart, the partners in the Biomimetic Competence Network and the related funding bodies and sponsors.

参考文献 / References：

[1] Terzidis, Kostas. Algorithmic Architecture. Oxford: Elsevier Architectural Press, 2006.

[2] Menges, Achim. "Material Computation: Higher Integration in Morphogenetic Design." Architectural Design 82 (2012): 14-21.

[3] Mueller, Gerd and Stuart A. Newmann. Origination of Organismal Form: Beyond the Gene in Developmental and Evolutionary Biology. Cambridge: MIT Press, 2003.

[4] Ball, Philip. "Pattern Formation in Nature: Physical Constraints And Self-Organiszing Characteristics.", Architectural Design 82 (2012): 22-27.

[5] Horowitz Frederick A. and Brenda Danilowitz. Josef Albers: To Open Eyes. New York: Phaidon, 2009.

[6] Otto, Frei and Bodo Rasch. Finding Form – Towards an Architecture of the Minimal. Stuttgart: Edition Axel Menges, 1996.

[7] Ahlquist, Sean and Achim Menges. "Computational Design Thinking," in Computational Design Thinking edited by Achim Menges and Sean Ahlquist, 10-29. London: John Wiley and Sons, 2011.

[8] Mitteroecker, Philipp and Simon Huttegger. "The Concept of Morphospaces in Evolutionary and Developmental Biology: Mathematics and Metaphors". Biological Theory 4 (2009): 54-67.

[9] Eble, J. Gunther. "Developmental and Non-Developmental Morphospaces in Evolutionary Biology." Santa Fe Institute Working Papers 99-04-027 (1999).

[10] Menges, Achim. "Morphospaces of Robotic Fabrication – From theoretical morphology to design computation and digital fabrication in architecture." in Proceedings of the Robots in Architecture Conference 2012 edited by Sigrid Brell Cokcan and Johannes Braumann, 22-47. Vienna, Springer, 2012.

[11] Menges, Achim, and Tobias Schwinn. "Manufacturing Reciprocities.", Architectural Design 82 (2012): 118-125.

Arts-based Robotic Applications through Dynamic Interfaces
From Fabrication to Performance in the Creative Industry
通过动态界面实现的机器人艺术应用
创新产业中从建造到表演

Johannes Braumann , Sigrid Brell-Cokcan / Association for Robots in Architecture & University of Art and Design Linz
约翰尼斯•布朗曼，西格丽德•布瑞尔-考根 / 建筑机器人协会 & 奥地利林茨艺术与设计大学

在过去的几年里，创意产业试图重新将数控机床（CNC）用于建造，并使其摆脱外包工厂的束缚。正因如此，定制化生产流程理念备受关注，使得即使不再是批量制造，仍旧可以确保高效地制造个性化产品。

就这点来说，3D打印机是最重要的工具之一。3D打印机从推出后备受欢迎，它的价格也随之大幅下跌。这确保了热衷于此的人们可以利用基于挤压成型的工艺来制造个性化产品。但是，这些机器的工作空间十分有限，并且它的工作循环速度非常慢。这是因为作为逐层增量制作的工艺，它要求前一层材料硬化以为下一层材料提供支持。因此，从打印小尺度的模型到打印足尺度的工业产品原型或者建筑构件，这需要巨大尺寸且极其昂贵的设备，而这种大型设备往往超出创意产业所能接受的范围。

而今，工业机器人可以实现超出桌面级CNC的加工尺度来填补这一空白。同时，它所能实现的并非只是建造，艺术家发现了6轴或多轴工业机械臂的动力美学，这使得工业机器人不仅仅用于建造领域；除了纯建造外，创意产业开始探索机器人的移动能力及其建造工艺所展现的艺术特征。

就其本身而言，数据交换是实现这些新的制造理念或动态性能的最大难题之一。尽管商用计算机辅助建造软件可控制机器人铣削及切割流程，但是，针对工业机器人研发的新的创新型工艺要求开发全新的建造策略或数据交互界面，以便将数据输入

In the past years the creative industries have attempted to take CNC (Computer Numerically Controlled) fabrication back into its own hands and away from outsourced factories. As such, an area of special interest is the idea of customized production processes, away from mass fabrication and towards individualized products that are still manufactured with great efficiency.

One of the most important tools in that regard are 3D printers, which have become highly popular and subsequently dropped steeply in price, making extrusion-based processes accessible to enthusiasts. However, the problem that these machines have in common that their workspace is significantly limited and that their cycle speed is inherently slow, being an additive process where material has to harden in order to support the next layer. Therefore the move from scale model to full-sized industrial-design prototypes or architectural building components required expensive, large-scale machines that were often out of the reach of the creative industries.

These days industrial robots have started to fill that void by enabling the creative industries to work on a scale that exceeds desktop-CNC machines. And this is not only done within the scope of fabrication – artists discovered the aesthetics of the kinematics of industrial robotic arms with six or more degrees of freedom. The creative industries start moving beyond pure fabrication to consider the movements of the robot and the process of fabrication as a performance and work of art in itself.

其他特定软件。以前，只有与工程师紧密合作才可能完成"创新性"机器人项目。但是，如今，新的工具，如KUKA|prc [1]及HAL[2]，使得建筑师、艺术家及设计师可自己定义及优化这些过程。

关于工业机器人

机械臂最先应用于汽车制造行业，其也广泛应用于点焊、夹板装载或其他需重复作业的产业中。机械臂主要包括6个独立的轴，并配有最小化传感器。其中，6个独立的轴产生了6个自由角度。最小化传感器通过探测过大扭矩或碰撞来保护机器人，并使得其比在美国国防部高级研究计划局发起的机器人挑战赛中所看到的自主式机器人更加简易。但是，在创意产业中，正是这种低复杂性使得机械臂备受关注，并且这种构造十分坚固可靠。在连续使用的情况下，平均故障间隔时间为8年；同时，还可以对其进行优化以便非专业操作员可进行操作。每年售出的工业机器人多达17万台 [3]。相对于其他的数控机床，如高端3D打印机及激光切割机的价格，机械臂的单独成本相对较低。此外，当有自动生产线关闭或升级时，仅用几千美元即可购买到众多二手工业机器人。

与常规机械相比，工业机器人最大的优点在于它固有的多功能性。机械臂可配备有类似于人手的基本工具，而无需使用数控路由器之类需要对其进行优化才能满足特定的任务要求。机械臂可利用其各个方位的工具来满足各种不同需求，而不会像三轴路由器那样仅能从上部切开。此外，可将机器人设置在工作区正中间而非工作区周围，由简易叉式升降机将其抬高。这既节省了空间，又确保了布置设备时能够具有更大的灵活性。

6轴机器的程序设计

所有上述特征使得工业机器人在创意产业领域备受欢迎。但是，当2007年开始利用工业机器人进行工作时，人们发现机器人工作流程的瓶颈是其程序设计——用于优化其工作的流程设计，也就是所谓的"教导"。比如，当机器人获得移动至某一位置的信息后，保存该位置信息，并最终重新按顺序进行操作。这种流程与自动化应用相关，当设定生产线后，可确保多年利用该生产线生产相同的部件。特别是创意产业要求设定确保有效制造原型产品的工作流程，这种流程使得可一次性制造这些原型产品而无需重复成千上万次。

为解决将"外部"数据转移到机器人上的问题，机器人生产商打造了新的界面。如通过像CAMRob或KUKA CNC，将G代码输入至机器人模拟环境，从而将CAM软件与机器人连接起来。但事实证明，这些环境对创意产业来说也是不理想的，因为它们极其昂贵，且依赖于同样昂贵的CAM软件，反过来其优化仅为减法生产创造工具路径。

当我们开始研究不基于工程软件的机器人编程环境时，

As such, one of the biggest challenges in realizing these new fabrication concepts or even kinematic performances lies in the exchange of data. While robotic milling and cutting processes can be controlled by commercial CAM (Computer Aided Manufacturing) software, new and innovative processes developed for industrial robots require the development of entirely new fabrication strategies or data interfaces to import data from other, specialized software. Where "creative" robotic projects were previously only possible in close cooperation with engineers, new tools such as KUKA|prc [1] and HAL[2] have given architects, artists, and designers the possibility to define and refine such processes themselves.

About Industrial Robots

Known primarily from the automotive industry, robotic arms are commonly used for spot-welding, palletizing, or other repetitive tasks. At a very basic level, they consist of six individual axes - resulting in six degrees of freedom - and are equipped with the barest minimum of sensors to protect the mechanics by detecting over-torque or collisions, making them far less complex than the autonomous robots that can be seen, for example, in the DARPA Robotics Challenges. However, for the creative industries, this low complexity is exactly what makes robotic arms so interesting: They are built to be extremely robust and reliable with a mean-time-between-failure of eight years of continuous use and are optimized to be handled by non-expert operators on the factory floor. As more than 170.000 industrial robots are sold each year [3], the individual costs of robotic arms are comparably low and fall below many other CNC machines such as high-end 3D printers and laser cutters. Furthermore, many used industrial robots are available for a few thousand US-Dollars, for example, when automotive production lines are closed or updated.

Their biggest advantage over conventional machines is their inherent multi-functionality. Rather than being optimized for a particular task such as CNC-routers, robotic arms can be equipped with basically any tool, similar to the human hand – and where a three-axis router can only cut from above, a robotic arm can apply its tools from all directions and orientations. Furthermore, instead of having to build a robot around its workspace, it can be placed right in the middle of it, carried by a simple forklift – generally saving space and allowing much greater flexibility in the placement of machines.

Programming Six-Axis Machines

All of these aspects make industrial robots highly desirable machines for the creative industries. However, when we started working with industrial robots in 2007, it quickly turned out that the bottleneck within robotic workflows is their programming, which is optimized for programming by demonstration – so called 'teaching' – where the robot gets moved to a position, and the position is saved and finally replayed in sequence. While such a process is relevant for

Grasshopper（一种十分便利的可视化编程环境）越来越受欢迎。通过结合所包含组件，我们可以创建一个参数定义，将工具路径转换为KUKA CAMRob可读取的5轴G代码；在定义后再进行进一步的强化，以直接书写KUKA机器人语言并包括正逆动态仿真方案，以预览机器人的运动。最后，我们还为Grasshopper创建了一系列机器人专用组件，以KUKA|prc的参数化机器人控制插件的名义进行了出版，该插件首次于ACADIA 2010上对外发布。

此后不久，我们成立了"建筑机器人协会"，旨在将工业机器人运用在创意产业中。一方面，我们不断追求这一目标，将协会作为创意产业机器人用户的网络，组织每两年一次的关于建筑、艺术及设计领域机器人制造的Rob|Arch会议，并在注重科学研究的施普林格出版社出版了相关的会议论文集[4, 5]；另一方面，协会同样是一个积极参与国内及国际研究项目的研究机构。最近，我们在林茨艺术与工业设计大学担任了创新机器人的教授职位，主要负责对工业设计学生进行机器人制造方面的入门教育。

在学术环境中教授机器人

在学术环境中教授机器人面对的主要挑战是我们按照创意产业要求开发定制软件的主要动力之一。那时，我们的机器人课程有60多名学生，课程内容不仅包括CAD软件，还包括CAM及机器人模拟环境。事实证明，课程量过大，每周几个小时的课程无法完成规定内容。

将机器人建造整合到创意产业的常用软件中，让我们可以教授学生多种技能；而且这种技能不仅仅适用于机器人建造，也适用于其他计算机数控或设计流程。我们不是为了让学生成为机器人处理及机器人控制的专家，而是希望通过仔细分析、评估、最终创建定制流程及工作流程让学生了解机器的运作。我们特别希望学生在一种我们称为"机器人建造内在设计"的流程设计中，可以考虑机器的性能。完成设计后，学生并不是开始创建制作流程，而是整合机器人在设计阶段的既有限制条件及可能性，以创建前所未有的高度优化和高效的流程。完成课程之后，学生应对机器人或一般机器流程的利弊有个大致了解，使他们之后在涉及设计变更的制造及评估时可作出明智的决定。

"机器人木工"项目：引入新的机器人方法

在"机器人木工"项目中，我们希望将创意产业的用户培养为机器人专家。该项目由奥地利研究基金的FWF'sPEEK项目设立，旨在提升基于艺术的研究。我们与维也纳应用艺术大学的艺术与技术学院一起申请了重负荷KUKA机器人装置，设在木工车间。在与工业设计师及木匠的合作下，研究项目旨在探索将机器人建造整合到传统木材施工工作流程中的新方法，或创造全新的流程。我们努力将机械知识整合到设计流程中。同时，

automotive applications where a production line is set up once and then keeps producing the same parts for years, the creative industries in particular need workflows that allow the efficient generation of prototypical objects that are only produced once, rather than thousands of times.

To solve the issue of getting data from "outside" to the robot, the robot manufacturers created new interfaces that would allow the import of generic G-code into the robot simulation environment, thereby linking CAM software to the robot, via CAMRob or KUKA CNC. However, even these environments proved not ideal for the creative industries, as they are highly expensive and rely on similarly expensive CAM software, which in turn is optimized solely for making tool paths for subtractive manufacturing.

The time when we started investigating robot programming environments that are not based on engineering software coincided with the growing popularity of Grasshopper, a very accessible visual programming environment. Just by combining the included components we were able to create a parametric definition that would convert toolpaths into five-axis G-code that could be read by KUKA CAMRob. This definition was later enhanced to directly write KUKA Robot Language and to include a forward and inverse kinematic solver to preview the robot's movements. Finally, we proceeded to create a range of dedicated robot components for Grasshopper that were published as KUKA|prc – parametric robot control and first presented at ACADIA 2010.

Shortly after we founded the Association for Robots in Architecture with the goal of making industrial robots accessible to the creative industries. We pursue this goal on the one hand as a network for robot users in the creative industries, organizing the bi-annual Rob|Arch conference on robotic fabrication in architecture, art, and design, and publishing the proceedings with the scientific publisher, Springer [4, 5], but also by acting as a research institution that is actively involved in both national as well as international research projects. Just recently, we have set up a professorship for creative robotics at the University for Arts and Design in Linz with the focus of exposing industrial design students to robotic fabrication for the very first time.

Teaching Robotics in an Academic Environment

The challenges of teaching robotics in an academic environment were one of our main motivations to develop custom software that was tailored towards the requirements of the creative industries. At that time we had robotic courses with more than 60 students and having to teach not only CAD software, but also CAM and robot simulation environments proved to be too much to cover in a few hours each week.

研究项目也将在设计早期整合更为复杂的木材性能，进行我们所说的"材料的内在设计"。在项目开始时，创意产业就表现出了极大的兴趣。木匠，甚至是学生都希望购买机械臂，以便在其设计中能够采用多轴铣削的方式进行操作。我们希望这些经济实惠的工业机器人与新的程序设计工具（如KUKA|prc）可以使小公司，甚至是个人在其日常工作中高效利用大规模的计算机数控建造。生产及材料的内在设计可确保高度优化的定制建造策略，而且具有当前工业流程所无法企及的高度灵活性。

AROSU项目：石材雕刻

传统机器人产业对于我们创造的高效工作方法同样觊觎，全球最大的航空制造商以及核工业制造商正利用我们的软件进行研究工作，并对复合材料部件进行非破坏性试验。这些应用都有行业标准计算机辅助建造软件未涵盖的特殊要求。AROSU项目（欧盟FP7计划赞助的研究项目）还碰到了另一行业挑战。天然石材雕刻数千年以来一直都是手动作业，由于缺乏专业石匠及该职业的健康风险越来越大，欧洲石材公司在大型项目中越来越难以与拥有廉价劳动力国家的公司抗争。通过定制机器人石材雕刻程序，AROSU项目研究小组希望一方面能够使石匠专注于专业性工作、降低健康风险；另一方面实施超越手工作业的新策略。在与研究合作伙伴多特蒙德工业大学、及商业合作伙伴Klero，KUKA，Bamberger Natursteinwerke，IIArchitects int和Gibson的合作下，正在研究开发针对石材雕刻的新软件和新硬件。

Integrating robotic fabrication into software that is commonly used in the creative industries now enables us to teach students skills that are not unique to robotic fabrication, but can also be applied to other CNC and/or design processes elsewhere. We do not want to make our students experts on robot handling and robot control, but rather expose them to the way machines work by closely analyzing, evaluating and finally creating custom processes and workflows. We especially want them to be able to consider the properties of the machine when they are designing – a process we refer to as robotic production immanent design. Instead of creating fabrication processes after the design, students integrate the constraints and possibilities of robots already during the design phase, leading to highly optimized and efficient processes that would not be possible otherwise. After a course, students should have an overview of the advantages and disadvantages of robotic or general machinic processes so that they can later on make an informed decision when it comes to fabrication and the evaluation of design variations.

Robotic Woodcraft: Introducing New Robotic Methods

A project where we do want to turn users from the creative industries into robot-experts is Robotic Woodcraft, funded by the Austrian Research Fund FWF's PEEK program to promote arts-based research. Together with the University of Applied Arts Vienna's Institute for Art and Technology we have applied for a heavy-payload

ADA项目：由康拉德·肖克罗斯制作的艺术装置，机器人编程由一个定制版本的KUKA|prc界面实现
The ADA Project: Robotic Art Installation by Conrad Shawcross, robot programming realized with a custom-tailored version of KUKA | prc

ADA项目：超越建造

在创意产业中，机械臂不仅仅用于建造，还用于动态安装。康拉德·肖克罗斯是一位居住在伦敦的艺术家，其已从事复杂动态安装十余年，精心制作的机械驱动钻机在受控的方式下移动灯光。为了实现其利用机械臂的愿望，我们与多学科团队（其中包括艺术家本人、专业的视觉特效/动画设计师大卫·弗拉姆霍克、及曾参与电影业数个创新机器人项目的机器人工程师凯文·阿莫斯）共同合作。动画设计师已在4D电影中开发出数个动态连接模型，意在打造能够将动画数据从4D电影导入到Rhinoceros/Grasshopper 3D中的定制工具箱，并可进行知识转化，使得科研团队能够快速编程（在4D电影中）、模拟（利用KUKA|prc）并进行机器人原型设计。

在4D电影中，动画设计师打造了虚拟动态连接，通过某一点进行驱动。连接中给出端点的运动在4D电影中是可追溯的，这使得在每一关键帧的X、Y、Z坐标内形成一点（如通常设置的24每秒帧数形成大约40ms的时间步序）。

但是，仅有工具作用点信息不足以确定机器人位置。机器人可通过无数种方式到达某一特定点，因此需要在Grasshopper内形成额外的数字工具，以动态调节工具矢量（即：方向），甚至围绕工具轴旋转。

一个重大的挑战是将时间因素应用到参数模型中。从理论上说，可以假设机器人从点A走到点B需要一个时间步序。点A和点B之间的距离已知，速度可以通过距离除以时间算得。这在纯数字模型中可行，但这并不能解释这一事实，即机器人的加速度不是无穷大的，因此机械臂需要一定的时间加速、制动至设定的速度。但有趣的是，高密度点也是另外一个问题。在2m/s的高速运动时，机器人1秒可以移动经过24个点位，相当于每80mm有一个点位。为了解决这一问题，可采取以下措施：调整机器人的内部插件，确保尽可能平稳地运动，并采用特殊的算法，可以减少类似的点位，如方向或速度上没有大变化的点位。

简化算法是根据定制的KUKA|prc CAM后处理程序改编的，在类似的问题出现时，输入G代码：若点密度过高，机器人速度放慢，会给表面饰面带来不利影响，如烧焦表面。在这种情况下，最好降低工具路径的分辨率，以便实现更准确的输出。

在不到一周的时间内，我们在伦敦的康拉德·肖克罗斯工作室，成功围绕现存的动画软件环境实施了一套工作流程，使得随后几年机器人安装舞台布景成为可能，如皇家歌剧院的芭蕾舞剧《提香》。

展望

今天，合理的价格以及可以大尺寸的加工使得机械臂对创意产业来说很有吸引力。无论是从建造还是从机器人性能上来说，

KUKA robot setup directly in the wood workshop. Together with industrial designers and artisans, the research project will explore new ways of how to integrate robotic fabrication into traditional wood-construction workflows, or to create entirely new processes. Similarly, just as we strive to integrate machinic knowledge in the design process, the research project will integrate the even more complex material properties of wood at an early design stage, towards material-immanent design. Even now at the very beginning of the project there is significant interest from the creative industries, with carpenters and even students looking to buy robotic arms as an affordable way to use multi-axis milling for their designs. We hope that affordable industrial robots, alongside new programming tools such as KUKA|prc will allow even small firms and individuals to efficiently use large-scale CNC fabrication in their everyday work. Production- as well as material-immanent design will allow highly optimized and at the same time customizable fabrication strategies with a flexibility that goes beyond current industrial processes.

AROSU: Structuring Stone

This efficiency is now also met with interest from the "traditional" robotic industry, with our software now being used for research at one of the largest global aircraft manufacturers, within nuclear plants, and for non-destructive testing of composite parts – each an application with specialized requirements that are not covered by industry-standard CAM software. Another industrial challenge is explored within AROSU, a research project funded by the European Union's FP7 framework. Structuring natural stone is a process that has been done manually for thousands of years, but the lack of expert masons and the increasing health risk for this profession are making it increasingly hard for European stone companies to compete in large-scale projects with low-wage countries. By creating custom processes for robotic stone structuring the AROSU research team is expecting on one hand to allow masons to focus on specialized work and detailing to reduce their health risk, while on the other side implementing new strategies that go beyond what is possible with manual processes. Together with research partners TU Dortmund and Labor, as well as commercial partners Klero, KUKA, Bamberger Natursteinwerke, IIArchitects Int and Gibson, new software as well as completely new hardware tools for stone structuring are being developed.

The ADA Project: Beyond Fabrication

Applications of robotic arms within the creative industries also go beyond fabrication towards kinematic installations. Conrad Shawcross is a London-based artist who has been working with complex kinematic installations for more than ten years, custom-building elaborate mechanical rigs to move lights in a very controlled way. To realize his vision of utilizing a robotic arm, we worked with an interdisciplinary team consisting of the artist himself, David Flamholc,

新的编程界面为富有创造力的用户提供了超出行业标准的建造方法。准确灵活地为机器人运动编程,可以通过将最初的用户转变成开发人员来实现,正如笔者和其他的一些机构,如密歇根大学[6]和斯图加特大学[7]创造了自己的、定制的工具包。通过这些工具包,我们就可以以最优化的参数化方式设计机器人的运动。

其中一些工具可供整个社区使用,使得用户可以利用现有的软件技巧,更快地控制这些复杂的、动态的机器。2006年,"控制"机器人需要花费几年的时间,今天,在研究实验室对机器进行设置后,很多新用户在几周内便可以输出成果。

有趣的是,现在,参数化机器人控制的知识与发明工业机器人的产业也密切相关,这使得我们可以在跨专业的团队中工作,将我们的知识与工业中机器人用户数十年的机器经验相融合。希望本文可以提供更有力、更易理解、更可靠的解决方案,为创意产业提供新的商业模式。

a professional VFX supervisor/motion designer, and Kevin Amos, a robot engineer who has been working on numerous creative robot projects in the film industry. As several kinematic linkage-models had already been digitally explored by the animator within Cinema 4D, the idea was to create a custom toolset capable of importing animation data from Cinema 4D into Rhinoceros/Grasshopper 3D and to initiate a knowledge transfer that would enable the team to quickly program (in Cinema 4D), simulate (using KUKA|prc) and prototype robotic choreographies.

Within Cinema 4D, the animator creates a virtual kinematic linkage, which is then actuated at a point. The resulting movement at a given endpoint of the linkage is traced within Cinema 4D, resulting in a point with *XYZ* coordinates at every key frame (commonly set to 24 fps which results in a time step of around 40 m/sec).

However, this information is not sufficient to uniquely define a robot position, containing only the information about the tool tip. As the robot can approach a single point in an infinite number of ways, additional digital tools had to be created within Grasshopper to dynamically adjust the tool vector (i.e. direction) and even the rotation around the tool axis.

A significant challenge was the implementation of the time factor inA significant challenge was the implementation of the time factor into the parametric model. Theoretically, it can be assumed that the robot needs one time step to go from one point A to the next point B. As the distance between A and B is known, the speed can be calculated easily by dividing the distance with the time. While this works in a purely digital model, it does not account for the fact that the acceleration of the robot is not infinite, so that the robotic arms requires a certain amount of time to accelerate and brake to the programmed speed. Interestingly, however, the high point density proved to be a more significant problem. At a high speed of 2 m/sec, the robot would have to move through 24 positions each second, which corresponds to one position every 80mm. To alleviate that problem, several measures were taken: the internal interpolation values of the robot were fine tuned to allow enable the smoothest-possible movement, and a special algorithm implemented that would be able to reduce similar points, when there aren't any significant changes in neither direction nor speed.

The reduction algorithm was adapted from a customized KUKA|prc CAM postprocessor that imports G-code as a similar problem applies: In the case of excessive points density, the robot slows down, which negatively impacts the surface-finish by charring the surface. In this cases it can therefore be preferable to reduce the resolution of a toolpath in order to achieve a more precise output.

Within less than a week at Conrad Shawcross' workshop in London we managed to implement a workflow around the existing animation software environment, enabling a number of robot installations over the following years, such as the ballet Titian at the Royal Opera House.

AROSU项目:高速图像采集和分析人工表皮结构是这一研究项目的部分组成。
High-speed image capture and analysis of manual surface structuring as part of the AROSU research project.

Outlook

Today, the affordable price and large size makes robotic arms very appealing to the creative industries. New programming interfaces even allow creative users to move beyond industry-standard fabrication methods and accurately and dynamically program the robots' movement, be it for fabrication or as a robotic performance.

This was made possible by turning initial users into developers. As such the authors, but also several other institutions like the University of Michigan [6] and the University of Stuttgart [7], created their own, customized toolsets through which they can plan the robot's movement in a highly optimized and parametric way.

Some of these tools are now available to the entire community, allowing users to quickly control these complex, kinematic machines with their existing software skills. Where in 2006 it took several years to get the robot "under control", many new users today are able to output the first results within a few weeks after setting up the machine in their research labs.

Interestingly, the knowledge of parametric robot control is now also becoming relevant to industries from which the industrial robots originated, allowing us to work in highly interdisciplinary teams towards integrating both our knowledge as well as the decades of machine-experience of the robot users in industry. Hopefully this discourse will lead to even more powerful, accessible, and reliable solutions and to new business models within the creative industry.

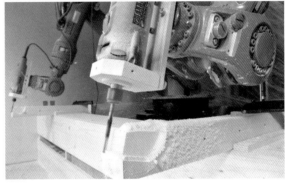

上图为技术人员在KUKA KR120 R2500 型号的机器人上安装一个定制的刀具，下图为两台KUKA KR16型号的机器人在铣削自定义的节点单元。
Artisan mounting a custom turning tool on a KUKA KR120 R2500 industrial robot (above), two KR16 robots milling a custom node element.

参考文献 / References :

[1] Braumann, J. and S. Brell-Cokcan. (2011). "Parametric Robot Control: Integrated CAD/CAM for Architectural Design." In Proceedings of the 31st Annual Conference of the Association for Computer Aided Design in Architecture (ACADIA), 242-251. Banff (Alberta).

[2] Schwartz, T. (2012). "HAL." In Rob|Arch 2012: Robotic Fabrication in Architecture, Art, and Design, edited by S. Brell-Cokcan and J. Braumann, 92-101. Vienna: Springer Verlag.

[3] IFR Statistical Department. (2014). World Robotics 2014 – Industrial Robots.

[4] Brell-Cokcan, S. and J. Braumann (eds.). (2012). Rob|Arch – Robotic Fabrication in Architecture, Art, and Design. Springer Verlag, Vienna.

[5] McGee, W. and M. Ponce de Leon (eds.). (2014). Rob|Arch 2014 – Robotic Fabrication in Architecture, Art, and Design. Springer Verlag, Vienna.

[6] McGee, W., Pigram, D., and M. Kaczynski. (2012). "Robotic reticulations: A method for the integration of multi-axis fabrication processes with algorithmic form-finding techniques" in Proceedings of the 17th International Conference on Computer Aided Architectural Design Research in Asia, 295–304. Chennai (India).

[7] Krieg, O., Dierichs, L., Reichert, S., Schwinn, T., and A. Menges. (2011). "Performative Architectural Morphology: Robotically manufactured biomimetic finger-joined plate structures" in Proceedings of the 29th eCAADe Conference, 573-580. Ljubljana (Slovenia).

METHODOLOGY
方法研究

Gramazio Kohler Research, ETH Zurich
苏黎世联邦理工学院"格拉马齐奥与科勒"研究中心

Robothouse, SCI-Arc
美国南加州建筑学院机器人实验室

ICD Institute for Computational Design, University of Stuttgart
德国斯图加特大学计算设计学院

Digital Design Research Center, CAUP, Tongji University
同济大学建筑与城规学院数字设计研究中心

New-Territories
新领域事务所

2013年，ETH与新加坡未来城市实验室合作的机器人建造高层建筑设计项目
Design of Robotic Labour High Rises, Gramazio Kohler Research, SEC Future Cities Laboratory and ETH Zürich, 2013

A New Building Culture
Towards a Radical Confrontation between Data and Physics
一种新的建筑文化
走向数据和物理间的激进对抗

Jan Willmann, Fabio Gramazio, Matthias Kohler / Gramazio Kohler Research, ETH Zurich
简·威尔曼，法比奥·格拉马齐奥，马蒂亚斯·科勒 / 苏黎世联邦理工学院 "格拉马齐奥与科勒" 研究中心

当今世界建立了一套完整的建筑技术基础体系，这从其发展伊始——20世纪初期建筑工业化就比现实更具想象力。我们见证的不再是现代化进程上的一次产业发展的滞后，相反，却是一种历史意义上的分离：脑力劳动与手工生产，设计与实现之间的现代主义意义层面的关系正被宣告结束[1]。于此同时，大量的建筑内在主体内容正在重新回归，不仅仅是手工工艺和建造的艺术，尤为突出的是建筑设计的方法。随着机器人在建筑学领域的普及，关于第二次数字时代的辩论题目不再是20世纪90年代由数字建筑提出的"从去物质化到获取纯粹形式"[2]。相反地，如今我们聚焦的是建筑广泛的数字化进程，这在其生存条件中包含着一个激进的范式转换。正如我们近期出版的新书《接触机器人——机器人如何改变建筑学》，在建筑领域对机器人的应用完全开辟了新材料的崭新愿景，而这一点从根本上改变了建筑设计和建筑文化[3]。

回顾历史

从20世纪90年代初期开始，从传统工业生产技术到数字建造流程的过渡转变引发了一场影响深远的变革，引领了建筑生产环境的范式革命。如今，非标准生产工艺已成为司空见惯的场景；过去，标准化一直是工业生产技术革新的推动力；而如今，所谓的"单件"生产——即生产某一特定商品——已然成为信息时代的推动力。由于工业机器人可以执行无限量的非重复任务，因此被认为是这一重大转变中的主要促成因素的其中之一。

Today, a uniform technological basis for architecture has been established, which from the onset of building industrialization in the early 20th century until now had been more a dream than reality. We are no longer witnessing the delayed modernization of an industry, but rather a historical departure: the modern division between intellectual work and manual production, between design and realization, is being rendered obsolete[1]. At the same time, a wide range of inherently architectural topics are finding their way back on to the agenda, not least among which are crafts and the art of construction, and, in particular, methods of architectural design. As robotics becomes increasingly commonplace in architecture, the subject of debate in the Second Digital Age can no longer be its "dematerialization into pure form", as had been proposed by digital architecture during the 1990s[2]. Instead, what we are observing today is the comprehensive digitalization of architecture, which entails a radical paradigm shift in its production conditions. As outlined in our recent book "The Robotic Touch – How Robots Change Architecture", the employment of robotics in architecture is thus opening up the prospect of entirely new material capacities that could fundamentally alter architectural design and the building culture at large[3].

Forward to History

Since the beginning of the 1990s, the transition from traditional industrial production techniques to digital fabrication processes

然而，相对专注于机器人本身的技术开发，不管其有多么迷人，我们更为感兴趣的是在建筑的视角下，探索潜在的机器人建筑设计方法及其实现过程。为此，我们又开始重新利用可靠、灵活的机械臂及经济实用的制造机械，因为这些工具的潜力早在传统工业应用中已得到充分验证。重要的是，建筑及其产品生产方式对机器人建造的方法有很强的启发性，反之则不然。如此，扩大建筑设计的范畴和生产的可能性才能促进新的材料分化及复杂性的涌现，并找到正确的实现方式。

实际上，20世纪90年代建筑建造工业对机器人系统的初步尝试绝对谈不上成功。首先，对于基于机器人的建造流程开发，其结果往往是既不专业又极其昂贵；同时，灵活性也十分有限。基于机器人的建设工厂同样也无法避免这些弊端。到目前为止，建筑施工中的机器人仅用于进一步优化（标准化）建筑流程，作为提高生产率的一种方法[4]。最后，没有真正持久的增加建筑价值，更别在将机器人整合到建筑产业的初步尝试中，会有任何新的（数字化）建筑文化生根发芽。但是这一切在千禧年到来之际发生了彻底改变，首先，数字技术的采用在建筑专业中越来越普遍，开始在对建筑设计及做法的理解方面有了更大的影响。此外，随着从其他产业借鉴的计算机控制生产机械的快速传播，如铣削、激光切割机或3D打印机，"数字项目"获得了相当的"材料价值"[5]。在2005年，为检查建筑的新生产条件，苏黎世联邦理工学院建立了一个采用工业机器人的多功能建造实验室——这是建筑领域的第一个机器人建造实验室[6]。

授权叠加的建构过程

实际上，采用机器人施工有多种可能性。机器人能够实施的空间物理操作范围几乎是无穷尽的。尽管机器人的操作受限于事先规定的运用范围，但机器人的"动手能力"可以自由地被设

建立于2005年的ETH机器人建造实验室，是建筑领域的第一个机器人建造实验室，它由一个能直线移动7m的Kuka KR 150 L110型机器人构成，这确保了能操作大型的建筑构件。
Robotic Fabrication Cell, Gramazio Kohler Research, ETH Zurich, 2005: The world's first fabrication facility for employing industrial robots in architecture – consisting of a Kuka KR 150 L110, which moves on a 7 metre long linear track and is thus able to assemble large architectural elements.

has triggered a far-reaching change, which has led to a paradigm shift in the production conditions of architecture. Now nonstandard manufacturing techniques have become commonplace. Just as standardization has been the driving force for technological innovation in industrial production in the past, so-called "oneof-akind" production – that is, the manufacture of unique pieces – functions as a driving force in the information age. The industrial robot, because of its ability to perform an unlimited variety of non-repetitive tasks, is considered as one of the key enablers for this deep transformation.

However, rather than focusing on the technological development of robots themselves, no matter how fascinating this might be, we are interested in establishing an architectural perspective on them by exploring the potential of robot-induced design- and materialization processes. To this end we have reverted to using articulated-arm robots as established, cost-efficient fabrication machines that are at once both reliable and flexible, and whose potential in conventional industrial applications has been thoroughly proven. It is essential that architecture and the conditions specific to its production inform the approach to robotic fabrication, and not vice versa. Only in this way it is possible to significantly expand the range of architectural design and production options, enabling a new material differentiation and complexity to emerge and find expression.

In fact, the early introduction of robotic systems to the building construction industry during the 1990s was anything but a success. Most of all, the development of robot-based construction processes frequently led to either highly specialized, extremely expensive construction robots with limited flexibility, or to robot-based construction factories yielding the same constraints. Robots in building construction up to this point were exclusively used to further optimize (standardized) building processes, as a means of achieving greater productivity [4]. Ultimately, no real lasting added architectural value, let alone any new (digital) building culture took hold during these initial attempts at integrating robotics into the building industry. All of this radically changed at the turn of the millennium. Digital technologies became more commonplace among the architectural discipline and began having a greater impact on the understanding of architectural design and practice. In addition, with the rapid spread of computer-controlled production machinery borrowed from other industries, such as milling and laser cutters or 3D-printers, the "digital project" attained considerable "material value" [5]. In 2005, in order to examine the resulting new production conditions for architecture, a multi-purpose fabrication laboratory employing an industrial robot was installed at ETH Zurich – the first such laboratory in the field of architecture [6].

Authored Additive Constructive Processes

The possibilities for implementing constructive processes with the robot are manifold. The robot can carry out a nearly unlimited

计和编程[7]。我们可以根据特定的材料和概念上的施工意图定制机器人的材料处理技能。与传统的施工流程相反，机器人操作并不能通过几何描述实现，而是需要通过编程的方式表达算法并加以记录。机器人加工过程信息包含了时间进度和建造顺序等多个参数，这些信息可以被机器人直接用于定义材料的空间拼接与生产过程。相比减法和叠加成型技术，这种授权叠加处理流程植入了更多的信息，允许利用基本材料聚集实现极其复杂的高性能建筑构件。此外，以这种方式设计的结构所需的材料比其他制作技术要少，但却能实现更具结构性能的建筑。

机器人对于建筑产业的深远影响在于设计和施工不再仅仅被定义为是相对独立的一系列实施阶段。相反，设计和施工在本质上互相交织，因为设计在最初启动时就已经包括了相关的机械建造能力，这就是机器人叠加建造的本质。高度可靠的建筑元素可以全尺寸组合建造；同时又允许本质上有差异的建筑风格，确保在特定的建造和结构条件下可以充分利用材料性能。

"冈特拜恩酒厂"项目的立面是第一个能够代表这种施工方式的建筑项目[8]。这个项目意义重大，因为它预示着机器人全尺寸叠加建造的原则，它证明了可以用非标准的方式装配无数个单元构件（砖块）。这种个性化的建造方式解释了为什么我们许多项目的建造基础只是简单的常见基本构件（如砖块或木板），但这些通用构件却能够组合成为高度科学、各式各样的建筑聚合物。在这个过程中，机器人的运用不仅意义重大，更是不可或缺的。相反，一旦单元构件通过几何相连而变得更加具体时，对单元构件的拼接很大程度上是预设的，施工灵活度因而受到限制，结果有时候通过人工组合这些构件反而比机器人更快。在这种情况下，机器人的特定附加值降低了，其意义仅仅在于人工流程的自动化操作。实际上，我们在苏黎世联邦理工学院的所有项目的特点，都是对大量建筑构件的极其具体

number of physical operations in space. Instead of being restricted in its operations to a prescribed range of applications, the "manual dexterity" of robots can be freely designed and programmed [7]. Their material manipulation skills can be customized to suit a specific constructive intention, both at the material and conceptual levels. In contrast to conventional construction procedures, however, the description of these operations can no longer be achieved by means of geometric depictions; rather they must be algorithmically denoted and "recorded" through programming. The fabrication instructions thereby produced – containing, for example, timing and building sequence – are directly used by the robot for the spatial joining of the materials. In comparison with subtractive and formative techniques, this shift towards authored additive processes allows for the aggregation of very complex and high-performing building components out of basic materials. Furthermore, the structures conceived in this way require less material than comparable fabrication techniques to derive similarly refined constructions, simply because the material is deposited at the exact location where it is

在"冈特拜恩酒厂"项目（2006）中的非标准砖墙立面中，这个400m²的立面完全是机器人通过对2万块砖块以不同角度旋转来砌筑实现的。因此，整体墙面的可塑性和由砖块不同旋转角度而带来的色调会因太阳的位置而变化。
Non-Standardised Brick Façade for the Gantenbein Vineyard, Gramazio & Kohler (in cooperation with Bearth & Deplazes), 2006: The 400-square-metre façade was robotically fabricated from 20,000 individually rotated bricks. As a result, the plasticity and hue of the rotated fields of bricks changes depending on the position of the sun, while from close up the three-dimensional depth dissolves and disappears in the detail of the bricks.

在2013年做的复杂木结构项目中，为机器人空间组装单元木构件而定制开发的空间连接节点系统，通过对于连接节点三角形形状排列的大小控制，可以使它的刚度能适应不同的具体的结构需求。在最终的1:1模型中根据误差补偿、结构性能和设计自由度来进行连接的可行性测试。
Complex Timber Structures, Gramazio Kohler Research, ETH Zurich, 2013: up: Custom-developed spatial node connection for spatial robotic assemblies – by increasing or decreasing the size of its triangular arrangement, the stiffness of the connection can be individually adapted to specific local structural requirements; down: final 1:1 prototype, testing the feasibility of the connection according to tolerance compensation, structural performance and design freedom.

的组织,这种组织方式使单元构件及其整体之间有鲜明的统一性和高度的清晰度。

机器人建造的新尺度

此后我们继续扩大了研究范围,通过一系列项目突出我们的研究重点。在这基础上,机器人流程的运用范围从预加工领域扩展到直接在施工现场上使用机器人和工业化全尺寸施工。例如,"序列屋顶"项目[9]就代表了我们的研究成果。该项目是用完全工业化制作的方式构建一个2 300m²的大型木屋顶,这是一个由48 000多根木杆以交替叠加的方式自动装配成的全尺寸建筑结构项目。与砖块相反,木材方便在加工过程中控制几何外形。机器人可以轻易将各个单元构件切割成特定的长度和角度。材料定制化使得这种流程成为可能。将通用的标准工业化产品转化为特定的机器人建筑构件,增加了施工体系内的灵活度。这种灵活度允许精巧结构的实现,促使了平面和曲面之间的无缝融合。

在"序列屋顶"这个项目里,离散的标准化木构件通过堆叠形成了一个特殊的连续渐变结构——在建造的概念上。一般性和特殊性的边界在这个结构中被模糊了,工业化和定制化的隔阂也消除了,因此产生了某种含糊的感知[10]。通过系统化的编程,结构体系能够针对材料和构件所在位置的特殊情况,灵活地做出构造上的回应,并可根据材料属性、施工和建造参数,整体地进行优化。这种复杂的系统显然无法通过传统的人工设计方法实现。例如,如果移动屋面结构中的一根木杆,其他无数关联构件的几何和建构关系都必须随之修改,单个构件和整体的关系也随之改变。当建筑单元构件的数量达到某种"临界值"时,数字化设计和建造就不再仅仅是有所帮助,而是成为不可缺少的一环。

实质上,建造逻辑与控制也将遭遇诸多挑战以及受到多重约束。对于建造而言,一方面,存在物质性约束(如载重量、加工零件的尺寸等);另一方面,还要考虑到经济性(如周期、返工率等)。数字设计必须考虑这些限制条件,并且在将设计数据转化为建造数据时也应考虑进去。换句话说,机器人建造程序必须采用有逻辑的和有效的顺序,以在整个建造过程中实现合理的周期。在这种情况下,我们无法再以静态的图纸去描绘设计,我们需要的是一整套基于编程的规则。建造和设计以这种形式无缝地连接在一起,即使是在设计的最后阶段。

如果建筑师意识到这种适用性并进行参数化,数字设计将变得格外有利。这种方式,并非只是纯粹功能主义范畴的操作,而更多地关注流程中的设计和建造方法,以期最终能够达到构造和材料应用上的创新。虽然"序列屋顶"项目将在2016年完工,但我们此前的研究已成功地展示了未来机器人建造将能达到的新尺度,并提供了具体的计算设计及建造方法,同时还可应用于该领域中的其他研究中。

actually needed for construction.

The farreaching consequence for architecture is that design and construction can no longer be defined as operatively separate, sequential phases. Design and construction are intrinsically interwoven as the design already contains the knowledge of its machinic constructability at the earliest point of its conception. Exactly here lies the essence of additive fabrication with the robot: highly informed building elements can be built up at full scale while being architecturally differentiated to their very core, enabling the full use of the material capacities within specific constructive and structural conditions.

The Gantenbein Vineyard Façade represents one of the first projects to demonstrate this immanent architectural potential [8]. It is remarkable because it anticipated the central principle of additive robotic fabrication at full scale by demonstrating the non-standardized assembly of an extraordinarily large number of single (brick) elements. With this characterization it becomes clear why many of our projects build on supposedly simple, that is, generic basic elements like bricks or timber slats. To the extent to which generic elements can be put together into various, highly informed and differentiated architectural assemblages, the application of the robot becomes not only meaningful but indispensable. Conversely, as soon as the individual elements become specific through geometrically prescribed connections, their joining is largely predetermined, and constructive freedom becomes limited. The consequence is that sometimes such elements would be put together more easily and perhaps also more quickly by hand than with the robot; in these cases, the specific added value of the robot would be reduced to the pure automation of manual work processes. Indeed, all of our projects at ETH Zurich are usually distinguished by a large number of elements and their very detailed organization, a high degree of definition throughout, and simultaneously a distinctive coherence between the single elements and the whole.

New Scales Of Robotic Fabrication

We have since further expanded our explorations and sharpened our focus through a number of research projects. As such, the range of robotic processes is gradually expanding, from prefabrication towards the direct use of robots on the construction site and constructions at full industrial scale. For example, The Sequential Roof [9] showcases this investigation. Conceived as a 2 300 square meter large timber roof, this project radically targets an industrial scale, where more than 48 000 alternately layered timber components are automatically assembled into a full-size building structure. In contrast to bricks, this material allows the manipulation of its geometry during the fabrication process. The robot can easily cut each element to a specific length at any

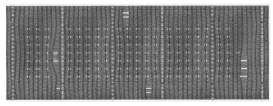

在"序列屋顶"项目中，48 000多根单独加工的木条以连续渐变的建造形式叠加，构成了2 300m²的屋顶。屋顶结构由168个跨度为14.7m，宽度为1.15m的梁格组成。每根木条的切割和定位都是不同的。每个梁格的加工由定制的6轴高架式机器人完成，它具有一个带有机械臂的平移轴，能够在48m x 6.1m x 1.9m的有效工作空间中进行三个旋转轴的全自动建造任务。这个屋顶是为在ETH未来的建筑技术实验楼设计的，它将在2016年完工。
The Sequential Roof, Gramazio Kohler Research, ETH Zurich, 2010-2016: Top view of the 2,300 square meter large "The Sequential Roof", assembled from 48,000 individually manipulated timber members to form a continuously graded constructive arrangement. The roof structure consists of 168 lattice girders with a regular span of 14.70 metres and width of 1.15 metres. The cutting and positioning of each timber element is unique. Custom six-axis overhead gantry robot of ERNE Holzbau AG, featuring a translational axis with a mechanical wrist and three additional rotational axes to perform fully automated fabrication tasks within an effective workspace of 48m x 6.1m x 1.9m. The roof is design for the future Arch_Tec_Lab building of the Institute of Technology in Architecture (ITA) at ETH Zurich that will be finished at 2016.

angle. The material customization enabled by this process – in which a generic, standardized industrial product is transformed into a specific and robust architectural element – results in additional degrees of freedom within the constructive system. This freedom allows for the realization of delicate structures in which plane surfaces can seamlessly merge with curved surfaces.

The explicit presence of The Sequential Roof results from the discrete layering of single timber elements in a continuously graded arrangement, which blurs the boundaries between the generic and the specific, the standardized and the individual, and therefore provokes a perceptive ambiguity [10]. Through programming the design is capable of responding to local requirements in a constructively flexible and specific way. By so linking the formal, constructive and fabrication parameters, an optimization of the entire structure can be achieved. Such systemic complexity can obviously not be managed by means of conventional manual design techniques. For instance, if only a single timber element of the roof construction is shifted, an endless number of relations in the complementary logic between geometry and tectonics, between the individual element and the entire structure, are changed. Here it becomes apparent that with a certain "critical mass" of construction components with mutual dependencies, the use of digital design and fabrication processes becomes not only meaningful but mandatory.

Essential here are also certain challenges with respect to fabrication logics and control, where multiple constraints evolve. On the matter of fabrication, on the one hand, these are physical constraints (e.g. loading capacity, dimensions of processed parts, etc.), whereas, on the other hand, these are also of economical nature (e.g. cycle time, reject rate, etc.). These constraints either have to be integrated already into the digital design process or have to be accounted when translating design data to fabrication data. Namely, the robotic fabrication process has to be structured into logic and efficient sequences, so that a reasonable cycle time can be achieved at full-scale fabrication. Correspondingly, the design can no longer be encapsulated in a static plan. Rather it is described by a programmed set of rules, which advantageously allow for seamless adaptations, throughout and even at very late stages of the design process.

This becomes particularly powerful if this adaptability is consciously anticipated and deliberately parameterized by the architect. The intent is thus less a purely functional modus operandi of construction optimization than a focus on equally processdriven and constructionaware design methods, ultimately to strengthen architecture in its constructiveinnovative character. Even though "The Sequential Roof" will only be completed with the new Arch_Tec_Lab building by 2016, the previous development has successfully demonstrated new dimensions of future robotic fabrication and provided a specific computational design and

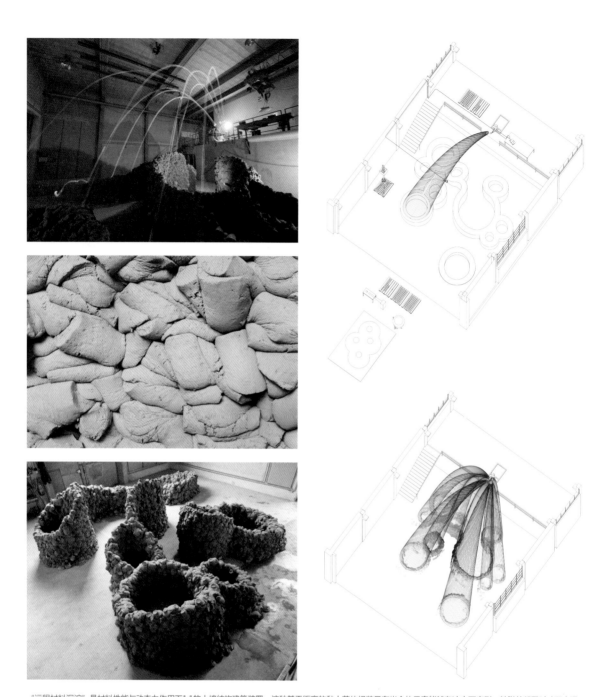

"远程材料沉淀"是材料性能与动态力作用下1:1的土墙结构建筑装置。这种基于距离的黏土落体组装只有当个体元素能够在冲击下变形、粘附并凝固时才可实现。在受冲击发生沉淀的瞬间，材料性能并不能完全体现，因此旋转的二维激光扫描仪可通过采样距离测量数据即时生成聚合流程的结构点云。最终生成的实时3D模型允许依据真实建筑状态对材料的模拟沉积进行调整，并确保最终的建造结构能够匹配定义的参考几何。这一研究从根本上扩展了机器人在预设几何的数字化建造之外的建筑能力，极大地为设计师提供了广阔的未探索的设计空间。

Remote Material Deposition, Gramazio Kohler Research, ETH Zurich, 2014: Remote Material Deposition was a 1:1 architectural installation demonstrating the design of loam wall structures between material behaviour and dynamic forces. This distance based assembly of loam projectiles can only be achieved when the individual elements are able to deform and adhere upon impact, and harden after. Since the material behaviour through the impact in the moment of its deposition cannot be predicted to its full extent, a rotating 2D-laser scanner was used to constantly generate a structured point cloud of the aggregation process through sampled distance measurement data. The resulting real-time 3D-model allowed adapting the simulated depositing of the material according to the actual building state and therefore to ensure that the fabricated structure ultimately matches the defined reference geometry. This radically extends the inherent architectural capabilities of robotic fabrication far beyond geometrically predetermined digital manufacturing and confronts the designer with a massively expanded and largely unexplored design space.

远离决定论：材料驱动下的建造过程

通过编程与施工过程直接对接，设计师可明确地接触现场施工过程。自此，计算机辅助设计媒介不仅为复杂而必要的设计系统带来机遇，如"序列屋顶"项目；同样可进行不确定的动态物质化流程设计，这一点在"远程材料沉淀"[11]项目中尤为明显。在该项目中，由机器人投掷塑性变形材料，并聚集成一个自密实结构。通过远程沉淀材料，机器人可克服传统数控生产机器的主要工作场所限[12]。正因如此，机器及加工品以类似"基于距离的3D打印"流程得到了分隔。

在"远程材料沉淀"项目中，每个投射物（平均重量1.8kg）的发射速度达到9m/s，满足之前设定1.8~14.0m（最大值）的目标。另一方面，材料的冲击位置及冲击行为都是可以准确预测的。因此，此类远距离集合的建筑结构设计是以投掷运动的方向、速度及顺序的定义为基础的。机器人投掷未固化材料的时间顺序将引起不同的结果。因此，设计不再是单纯以几何结构为基础[13]。可以说它是一个整合特殊材料特性的反馈过程。这促使了整个过程从确定的机器人建造流程转向自适应的建造过程。为了真正意义上的设计并且控制这些建造过程，必须通过传感器控制的反馈系统来实时掌握实际材料建造状态的信息。关于每个聚集状态的反馈信息使适应于实际建造物的抛射在不断调整中成为了可能。

因此，设计过程变成了材料特性及动态力学[14]之间的"建构校准"。对投掷曲线、冲击情况及反馈过程的定义成为了建筑设计中一个不可或缺的部分，同时将机器人建造的固有建筑能力延伸至几何预定的数字建造之外。

机器人版本的新城市类型学研究

与数字化建造极端扩张的物质性相反，从属于新加坡未来城市实验室的"机器人建造高层建筑设计工作室"致力于研究机器人建造如何解决创新设计方法的问题[15]。在此实验装置中，机器人将计算机辅助设计与实体研究模型制造连接起来，以便通过数字-物理相结合的方法，使得实体模型在计算机时代也能重新获得重大意义[16]。除此之外，机器人建造高层建筑的设计项目仅仅是在大规模住宅塔楼开发环境下进行的对机器人建造探索的首次设计探索。为了克服如今城市开发中单调、重复的功能，项目的重中之重是通过数字设计和建造流程对创新建筑类型进行调查研究。

该工作室为了解决了这一问题，通过对一系列定制机器人的建造流程及其建造成果进行研究。针对城市品质、空间和功能多样性方面进行了深入的调查和相关可能性探索，还建立了专门的机器人实验室。在这里，50层楼高的1∶50的研究模型和参数化设计的综合性高层建筑均可通过机器人直接建造。利用定制的机器人工具包操控单个数字化建筑组件，通过非标准的组装建立模型，这些模型不仅具有特殊性，还可以进行反复的优

fabrication approach that can be transferred to other explorations in this field.

Turning away from Determinism: Material-Driven Fabrication Processes

Through the unmediated linkage of programming and construction the designer gains explicit and machinic access to the physical construction process. Here, the medium of computational design not only provides the opportunity to design systems that are complex by nature yet essentially constructively determined, such as, for example, "The Sequential Roof", but also allows for the design of indeterminate, dynamic materialization processes. This becomes impressively apparent in the project Remote Material Deposition [11], where plastically deformable material is thrown by a robot and aggregates into a self-compacting structure. On the one hand, by remotely depositing material over distance, the robot overcomes the prevailing workspace boundaries of conventional digitally-controlled production machines [12]. As such, machine and artifact are separated in a quasi "distance-based 3D-printing" procedure.

In Remote Material Deposition the launching speed of each projectile (average weight 1.8 kg) reached up to 9 m/s and allowed meeting predefined targets in a range between 1.8 m and a maximum of 14.0 m. On the other hand, neither the impact location nor the impact behavior of the material is exactly predictable. Consequently, the design of such a remotely aggregated architectural structure is based on the definition of the direction, speed and sequence of the throwing motions. The chronological sequence in which the robot throws the still uncured material leads to different outcomes. Accordingly, the design is no longer purely geometry-based [13]. Rather it is determined as a set of feedback processes that incorporate the specific material behavior.

This leads away from deterministic robotic fabrication routines towards adaptive constructive processes. In order to actually design and control these, however, the fabrication processes must include a sensor-controlled feedback system, in which the information about the actual material buildup state is made available to the designer in real-time. The feedback information about each state of aggregation makes it possible, then, to constantly adapt the throwing to the actual built reality.

The design process therefore turns into a "constructive calibration" between material behavior and dynamic forces [14]. The definitions of throwing curves, impact scenarios and feedback processes become an integral part of the architectural design and extend the inherent architectural capabilities of robotic fabrication far beyond geometrically predetermined digital manufacturing.

化，并根据实际建造进行修正。

同时，工作室的研究脱离了3D打印建筑模型通常会采用的固有模式，转而重视其内在的构造逻辑[17]。程序化的构造原则是指导机器人建造模型的关键。这对受制于实体构造原则限制的模型来说是非常重要的。他们不仅仅要作为承重件，而且要将建造和结构的逻辑同时引入混凝土模板。通过计算机辅助设计和机器人建造可以预测建筑流程，也可以说，通过材料试验，将构造特性融入程序化设计将意义深远。该研究模型必然会体现增量建构逻辑知识，因此，它们一定是真实的试验，可以在实体建构层面实现数字化设计，并将其表达出来。（关于该项目的详细介绍见本书P160~P165）

理想与现实间的激辩

"飞行组装建筑"项目采用了最激进的实现方式[18]。该项目在位于奥尔良的FRAC艺术中心展出，是首次使用飞行机器人进行安装的建筑项目——多个四角直升飞行机器人组装了1 500多个构件，打造出了多孔塔状聚合体[19]。

这些聚合体呈现了一个实际高度为600m，一座"垂直村落"[20]的1：100共180层的建筑模型。该城市结构可为三万居民提供居住空间。此结构利用网格状组织，赋予多功能模块以多样化布局，并实现了更大程度的自由。然而这种网格并未采用传统城市组织中的横向布置，而是采用了竖向布置；最终，网格的两端闭合起来，形成一个环形整体。因此，营造出的圆柱形结构不仅是一个自稳定结构，还体现了新型的空间分化。这是通过编程实现的多样性以及高密度的城市组织。

"飞行组装建筑"项目首先意味着技术的进步，展示了数字化建造规模的极速扩大。飞行机器人被蜂群般部署，在实际操作中，它们相互调整运行方式，合作完成"垂直村落"的建造工作。相应地（在此案例中）"四翼直升飞行机器人"被视作支撑技术，但它同时也发挥了开创新时代的作用，使得一种新乌托邦构想成为可能。

在这个项目上，空域不仅是一种施工环境，更成为了一种综合设计研究范式[21]。这就带来了一种"暧昧"的对比：来自于几乎就在参观者头上进行的、非常物理的建筑建造；以及其传递的是一种非常理想的、乌托邦式的理想城市的传统。[22]因此，"飞行组装建筑"项目促进了城市向机器人建造的新都市风格转变。在比实际小100倍的比例下，（机器人）安装工作动摇了理想和现实之间的那条本该清晰的界限。同时，建筑和机器人之间的界限也变得逐渐模糊，以至于"想象和实施之间的界限"被重新探索。因而，"飞行组装建筑"项目扩大了概念范围及机器人建造的规模。只有通过探索实体建造的可能性及其技术限制，才可以展现未来机器人建造的能力。

Physical Versioning and New Urban Typologies

In contrast to such a radically expanded materiality of digital fabrication, the Design of Robotic Fabricated High Rises studio at the Future Cities Laboratory in Singapore focuses on the question of how robotic construction can address novel design methods [15]. In this experimental setup, the robot links computational design to the fabrication of physical study models, so that through such hybrid digital-physical methods, the physical model – even in the age of computation – again gains central significance [16]. Beyond this, Design of Robotic Fabricated High Rises is surely one of the first design attempts to explore robotic fabrication in the context of large-scale residential tower developments. In order to overcome the prevailing paradigm of repetition and mono-functionality in such urban developments as well as the resulting monotony, the central concern of the project is the investigation into innovative building typologies through digital design and fabrication processes.

The studio addresses this by investigating a spectrum of bespoke robotic building processes and their architectural consequences in terms of urban qualities as well as spatial and programmatic diversity. In order to conduct the studio research and to explore these potentials, a dedicated robotic laboratory was set up. Here, 1:50 study models of parametrically designed mixed-use high-rise structures of up to fifty floors can be fabricated directly with the robot. Through the non-standard assembly of individual, digitally prefabricated building components through bespoke robotic end-effectors these models are not only exceptional in their differentiation but can also be iteratively refined and physically "versioned".

At the same time the studio turns away from the predominant representational mode of 3D-printed architectural models, where the underlying constructive logic of the design is often not embedded [17]. Here, in contrast, programmed constructive principles are pivotal to guide the robotic fabrication of the models. This is significant in that the models are subject to the constraints of tangible constructive principles; they have to function as physically load-bearing artifacts, and bring fabrication and structural logics into concrete form. Conversely, it is possible by means of computational design and robotic fabrication to anticipate the building process, so to speak, through material experimentation, and thereby to integrate its constructive nature into the programmed design. The study models embody in an inescapable way the knowledge of their own incremental building logic. They are thereby insistently real experiments that materially condition the digital design and bring it to expression. (See the detail of this project in P160~P165)

迈进数字建筑文化

诚然，现今的形势十分独特：机器人建造提供的机遇大大丰富了各专业的材料做法，并从内而外更新了建筑建造知识。在这种背景下，机器人建造方法是基于材料处理和其自身的构造能力所生成的建造行为，而不是将材料理解为仅为预定建筑外形服务的建造行为。这种方法在"冈特拜恩酒厂"项目的非标准化砖块立面、"序列屋顶"、"远程材料沉淀"和"飞行组装建筑"等项目中得到了证实——通过砖块的层次布置或独立切割木材构件、可变形的黏土投射物远程材料沉积或飞行实现施工。通过这些项目，我们旨在表达这一观点：机器人可以作为催化剂，来影响数字化建筑的文化意义。事实上，机器人使建筑的建造条件发生了深远变化，计算机设计与现实之间建立了亲密（创意）对话。机器人在这一过程中起着决定性作用，因为正是通过机器人才使建筑从数字变为实体及有形空间[23]。该做法还从建筑数字化中去掉抽象及强制的人为特性，赋予其独特的美学意义及识别性。这种"机器人应用"也由此强化了建筑的内在竞争力，在项目规划及现实之间建立了统一的联系。这样做的结果是建筑将不再依赖于外部的多样及复杂性来增强其美感，而是通过数字化建造的创意、材料及文化表达等内部表现来丰富建筑的美感，从而营造出了新的当代数字化建筑文化[24]。

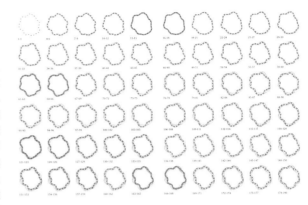

与传统的建造方式不同的是，在2011年的"飞行组装建筑"项目中，飞行机器人可以在空中自由操作，对它们要搭建的建筑材料进行临时组装。而由它们打造出来的多孔塔状聚合体所形成的"垂直村庄"为不同的城市项目提供了可能，25个单独定位的模块在每一层都互相作用，而各模块之间留有的空间也是不同的，但却形成了一个均匀的空间序列。有了这样一个由相关模块及模块之间的空间和连接所形成的网络系统，使得"垂直村庄"拥有了复杂分层的私密、半私密和开放的公共空间。这种分散化的布局避免了传统城市规划点状布局的限制以及现代城市通道拥挤的问题。

Flight Assembled Architecture, Gramazio & Kohler and Raffaello D'Andrea in cooperation with ETH Zurich, FRAC Centre Orléans, 2011: In contrast to conventional building processes, here the flying robots were able to operate freely in airspace and amalgamate temporarily with the building materials they ultimately deposit. The Vertical Village , formed by the porous, vertical (urban) aggregation, enables a varied urban program. Up to 25 individually positioned modules on each horizontal layer interact with each other. The areas in between vary and yet nevertheless form a homogeneous sequence of spaces. With such a network of interrelated modules, in-between spaces, and connections, the Vertical Village is formed by an intricate layering of private, semiprivate, and public space, fostering a decentralization that avoids not only the point-like restrictions of older urban planning and the gridlocked pathways of the modern city.

Negotiating between Utopia And Reality

This becomes clear in the most radical way in the project Flight Assembled Architecture [18]. The project at the Fonds régional d'art contemporain (FRAC Centre) in Orléans represents the first architectural installation to be built autonomously by flying robots: several "Quadrocopters" assembled over 1,500 elements to create a porous, tower-like aggregation [19].

What unfolds is a 1:100 scale model of an architectural vision for a 600 m high Vertical Village [20]. With a total of 180 floors, the urban structure provides living space for 30 000 inhabitants. The structure makes use of a grid-like organization that allows for a great degree of freedom in the variable arrangement of multi-functional modules. However, the grid does not run horizontally as in a conventional urban organization; rather it is turned vertically. Finally, both ends of the grid close to create a circular whole. Thereby the resulting cylindrical structure is not only self-stabilizing but also embodies a new type of spatially differentiated and programmatically diverse, dense urban organization.

Flight Assembled Architecture represents, above all, a technological intensification that introduces a radically expanded scope of digital fabrication. This results from the swarm-like deployment of the flying robots, in which they reciprocally adjust their operations and cooperatively build the Vertical Village. Correspondingly, the "Quadrocopter" is used as an enabling technology and also serves as a "conceptual door opener" in order to make possible a radical architectural utopia that does not need to exclude its concrete material implementation.

In this regard, the airspace not only corresponds to a constructive environment but also becomes a comprehensive design paradigm [21]. A sort of "suspense" results: between a physical architectural installation, which is being built almost over the visitor's head, and its manifestation as a future utopia in the tradition of ideal cities [22]. Flight Assembled Architecture therefore fosters a revision of the city as a robot-built urbanity. At a scale 100 times smaller than the vision, the installation calls into question the supposedly clear line between utopia and reality. At the same time the boundaries between architecture and robotics become less distinct here, so that the "border between the thinkable and the feasible" is newly explored. Flight Assembled Architecture thereby expands the conceptual scope and constructive scale of robotic fabrication. Only by exploring the physical possibilities and the specific limitations of this technology will the architectural capacities of future robotic fabrication unfold.

Towards A Digital Building Culture

The situation today is Truly unique. Robotic fabrication provides the opportunity to expand substantially the material practice of the discipline and to renew the constructive knowledge of architecture from the inside out. In this context, such a construction-related approach understands fabrication with the robot as something that emerges from the treatment of material and the understanding of its constructive capacities, rather than material merely serving a predetermined form. Projects such as the Non-Standardized Brick Façade for the Gantenbein Vineyard, The Sequential Roof, Remote Material Deposition or Flight Assembled Architecture demonstrate this approach, be it through the layering of bricks or individually cut timber elements, the remote material deposition of deformable loam projectiles or through airborne construction. With such projects we mean to convey that the robot can act as a catalyst to impart cultural significance to digital architecture. In fact, it characterizes a seminal change in the production conditions of architecture, placing computational design in a close (creative) dialogue with reality. The robot can thus play a decisive role specifically because through it, the digitalization of architecture becomes physical and tangible [23]. This takes away the abstract and forced artificial character from the digital in architecture and imbues it with a distinct aesthetic significance and identity. Such a "Robotic Touch" thereby strengthens the inner competence of architecture, providing a consistent bridge between the planning of a project and its realization. The result is an increased sensuousness that no longer originates from outside, through the favoring of formal exuberance and complexity, but rather from inside through the intensification of the creative, material and cultural expressive content of digitally fabricated architecture. And out of this grows a new contemporary digital building culture [24].

参考文献 / References:

[1] M. Carpo, Revolutions: Some New Technologies in Search of an Author, in: Log 15, New York (2009): 49–54.

[2] N. Kuhnert and A.-L. Ngo, Entwerfen im digitalen Zeitalter, in Archplus 189, Aachen/Berlin (2008): 7–9.

[3] F. Gramazio, M. Kohler, and J. Willmann (eds.), The Robotic Touch – How Robots Change Architecture, Zurich: Park Books, 2014.

[4] T. Bock, T. Linner, W. Lauer, and N. Eibisch, Automatisierung und Robotik im Bauen, in Archplus 198/199 (2010): 34–39.

[5] Cf. A. Picon, Digital Culture in Architecture, Basel: Birkhäuser, 2010.

[6] F. Gramazio and M. Kohler, Digital Materiality in Architecture, Baden: Lars Müller Publishers (2008): 7–11.

[7] T. Bonwetsch, F. Gramazio, and M. Kohler, Towards a Bespoke Building Process, in Bob Sheil, ed., Manufacturing the Bespoke, Chichester: John Wiley & Sons Ltd (2012): 78–87.

[8] "Non-standard Brick Façade for the Gantenbein Vineyard" [Online]. Available: http://gramaziokohler.arch.ethz.ch/web/e/forschung/52.html

[9] "The Sequential Roof" [Online]. Available: http://gramaziokohler.arch.ethz.ch/web/e/forschung/201.html

[10] Cf. J. Willmann, F. Gramazio, and M. Kohler, Die Operationalität von Daten und Material im Digitalen Zeitalter, in Positionen zur Zukunft des Bauens, Munich: DETAIL (2011): 6–19, .

[11] "Remote Material Deposition" [Online]. Available: http://gramaziokohler.arch.ethz.ch/web/e/lehre/276.html

[12] K. Doerfler, S. Ernst, J. Willmann, L. Piskorec, V. Helm, F. Gramazio, and M. Kohler, Remote Material Deposition, Proceedings of the COAC, ETSAB, ETSAV International Conference What's the Matter – Materiality and Materialism at the Age of Computation, Barcelona (2014): 101–117.

[13] K. Dierichs and A. Menges, Natural Aggregation Processes as Models for Architectural Material Systems, in: Proceedings of the Design and Nature Conference, Pisa and Southampton, 2010.

[14] F. Wittel, Single particle fragmentation in ultrasound assisted impact comminution, in: Granular Matter 12/4 (2010): 447–455.

[15] "The Design of Robotic Fabricated High Rises" [Online]. Available: http://gramaziokohler.arch.ethz.ch/web/e/lehre/219.html

[16] M. Budig, W. Lauer, J. Lim, and R. Petrovic, Design of Robotic Fabricated High Rises, in W. McGee and M. Ponce de Leon (eds.), Robotic Fabrication in Architecture, Art and Design 2014, New York: Springer (2014): 111–130.

[17] "The richness of this [constructive] repertoire contrasts with the simplicity of current 3D printing processes, which are limited to the two dimensional layering of material at a fixed resolution. Speed can be increased, material properties tuned, and the worry about an impoverishment of the architectural discipline dismissed as nostalgic, but the generic character of 3D printing represents a severe limitation to its application at the building scale." F. Gramazio, M. Kohler, and M. Budig, The Tectonics of 3D Printed Architecture, in: FCL Gazette, Issue 19, Singapore, 2013.

[18] "Flight Assembled Architecture" [Online]. Available: http://gramaziokohler.arch.ethz.ch/web/e/forschung/209.html

[19] F. Augugliaro, S. Lupashin, M. Hamer, C. Male, M. Hehn, M. Mueller, J. Willmann, F. Gramazio, M. Kohler, and R. D'Andrea, The Flight Assembled Architecture Installation: Flying Machines Cooperatively Constructing Structures, in: IEEE Control Systems Magazine, Volume 34, Issue 4 (2014): 46–64.

[20] J. Willmann, F. Gramazio, and M. Kohler, The Vertical Village, in: F. Gramazio, M. Kohler, and R. D'Andrea (eds.), Flight Assembled Architecture, Orléans: editions hyx (2012): 13–23.

[21] M. Kohler, Aerial Architecture, in: LOG 25, New York (2012): 23–30.

[22] A. Tönnesmann, Monopoly. Das Spiel, die Stadt und das Glück, Berlin: Verlag Klaus Wagenbach (2011): 85–126.

[23] F. Gramazio and M. Kohler, Made by Robots: Challenging Architecture at a Larger Scale, Architectural Design, Issue 229, London: John Wiley & Sons, 2014.

[24] This text is based on the author's publication: The Robotic Touch – How Robots Change Architecture, Zurich: Park Books, 2014.

在南加州建筑学院的机器人实验室中，6个工业机器人的重叠的工作范围
Overlapping work-spheres of the six industrial robots at the Robot House

A Creative Platform
一个创新的平台

Curime Batliner, Michael Jake Newsum, M.Casey Rehm / Southern California Institute of Architecture (SCI-Arc)
克瑞姆·巴特莱纳，迈克尔·杰克·纽斯曼，M. 凯西·莱姆 / 美国南加州建筑学院

背景

在过去十年间，机器人在建筑领域的应用已成为一种全球趋势。这正慢慢改变我们作为建筑师思考及工作的方式。借助于新的定制工具和流程，整个建筑领域的从业者正投入到这种新的工作方式中。正如法比奥·格拉马齐奥和马蒂亚斯·科勒所述："机器人在建筑领域的应用有潜力将整个行业重塑为一种实践；智能操作和人工生产，设计和实施之间的现代分工正逐步被淘汰。" [1]

尽管绝大部分措施直接针对建筑行业，但是全球金融危机，特别是在美国加利福尼亚州建筑公司业务惨淡的同时也使得机器人系统价格下降，进而激发人们产生采用不同的方法将机器人应用到建筑领域的想法，这对于重新设定工具和制定机器人加工流程有了独特的要求 [2]。

2011年，美国南加州建筑学院创立了机器人实验室。该实验室是一个由多机器人组成的设计实验室，旨在研究机器人同步协作及非线性建造流程。机器人实验室采用这种复杂的场景"来质疑建造本身及其逻辑结构的协议"以便产生新的建筑设计协议 [3]。它更关注于创建一个思想的对话与探寻，通过机器人诱导性设计和物质化流程来质疑当代建筑文化的创生，并创建一种新的建筑视

Context

Over the last decade, robotics in architecture have become a global trend which is slowly transforming the way we think and work as architects. Armed with new custom developed tools and processes, practitioners from the entire spectrum of architecture engage in this new way of working. As Gramazio & Kohler identify, "robotics in architecture have the potential to recast the entire field as a practice; the modern division between intellectual work and manual production, between design and realization, is being rendered obsolete." [1]

While most of the initiatives target the building industry directly, the context of a global financial crisis, which in the United States and in particular in California, has left architecture practices without work and simultaneously dropped pricing for robotic systems, stimulated the idea of taking a distinct approach to robotics in architecture which required its own alternative process of retooling and resetting [2].

In 2011, SCI-Arc opened the Robothouse, a multi-robot design facility researching collaborative synchronous robotics and

角，而不是直接针对建筑行业进行新的建造协议的开发。南加州建筑学院的机器人实验室关注于创建一种新的文化，通过对新一代设计师的培养，致力于通过建筑的对话参与到文化生产中。

理念

自创建以来，机器人实验室设备就被视为一个"图像诱导性机器"[4]。其源于建筑历史，在建筑历史中建筑表现和技术图纸一直是推动建筑发展的驱动力。在《自主转译》一文中，彼得·泰斯特写道："'房屋'的意义，从包豪斯到机器人工厂的转变体现了从机械形态主义到基于产品的协同共创衍生主义，从对现代机械的痴迷到向物品制造的转变"[5]。阿密特·沃尔夫解释到："在此可以提出一个形式理论，该理论直接针对服务特征或建筑同步生产……，并促进可能偏离现有建造理念，并反过来将体现真正的建筑物质性及可建造性的理念的传播。"[6] 同时，物品制造几乎是全球所有机器人项目的核心目标。南加州建筑学院的生产模式本身并不在用于批量制造的生产流程或实体对象范围内；相反，其注重于思想的创造；更确切地讲，其注重于"思想图像"的制造[7]。通过对所制造的物体的重新思考，我们注重于形象思维和非语言技术[8]。这与彼得·斯洛特戴克的论点是一致的，即，如果如今最先进的思想旨在"通过利用一种新的意识形式来描绘思想过程及数据格局"，那么久必须通过"图像表达"来实现[9]。

建筑本身就是一门混合艺术。在混合艺术中，图像不可脱离相关活动而单独存在。
Architecture itself,…, is a hybrid art, where the image hardly ever exists without combined activity. [10]

——伯纳德·屈米 Bernard Tschumi (1995)

在2012年的"热网"项目中，机器人需要处理大量装配序列
Hot Networks, 2012: Large Assembly Sequence

nonlinear fabrication processes. The Robothouse uses this complex scenario "to interrogate the protocols of manufacturing itself and its logical structure" with the goal to spin-off new design protocols in architecture [3]. Rather than developing new fabrication protocols targeting directly the building industry, it set out to create a discourse, a search for ideas, using robot-induced design and materialization processes to question the contemporary production of culture and establish an architectural perspective. SCI-Arc's Robothouse focuses on the production of culture, by casting a new generation of designers, aspiring to participate in fields of cultural production through the discourse of architecture.

Concept

From its inception, the Robothouse experiment was conceived as an "image-inducing machine" taking momentum from the history of architecture where representation and technical drawings have been driving forces in the development of architecture. [4] In the article "*Autonomous Translation*," Peter Testa writes, "undergrounding the 'house', from Bauhaus to Bothouse shifts from mechanomorphism to looser, extended morphisms based in production, from modernistic mechanic obsessions to the making of things" [5]. Amit Wolf explains that, "it is possible to speak here of a theory of form that directly tackles the attendant features or architecture's synchronic production…. and promotes relays that might deviate from established ideas of making and might in turn translate into genuine moments of architecture materiality and constructability." [6] Meanwhile, the making of things is at the core of almost all robotics initiatives around the around the globe, SCI-Arc's production mode does not constitute itself in a production process or physical object ready for mass fabrication. Instead, it focuses on the making of ideas, and, more specifically, the making of "images that mean ideas" [7]. By rethinking the manufactured object, it is argued that our interests focus on visual thinking and nonlinguistic technologies [8]. This is consistent with Sloterdijk's argument that most advanced thinking today must make use of "image giving" techniques if it is to "illuminate the landscapes of ideas, discourse and data with a new kind of conscious formal seeing." [9]

Platform

The initial three years of the work at the SCI-Arc Robothouse have been an intensive time of design explorations and ideations fueled by this conceptual framework. Software control models, which form central nodes in these networks of translation, facilitate the Robothouse's discourse. The hardware setups, custom software and particular ways of viewing the space, form what we refer to as the platform or the digital/physical design interface in the Robothouse.

Themes such as time, simulation, emulation and post-animation

平台

南加州建筑学院机器人实验室项目开始的三年主要是进行由这种理念框架推动的设计探索及构思。软件控制模型是构成这些转化网络的核心，用于帮助实现机器人实验室的项目。硬件装置、定制软件及诠释空间的独特方式构成我们所谓机器人实验室的平台或数字/物理设计界面。

时间、模拟、仿真及后动画这些主题是大部分工作的核心。事实证明这些方法对思想发展以及替代技能组合的发展有着巨大的积极潜力，它们同样也是各类探讨的核心话题，学院投票中的类似倡议也赞成采取更直接的方法生产设计物品。关于此背景下机器人应用的可扩展性问题是非常重要的，因为这迫使人们对试验做出批评性评估。

新一代教师如克瑞姆·巴特莱纳，迈克尔·杰克·纽斯曼和M.凯西·莱姆于近期加入了南加州建筑学院，并在此讨论中提出了新的视角。工作室及研讨会中的各种观察强化了这一想法，即在大量现实条件下以及实施过程中具有高度交互性的项目开始蓬勃发展。他们提出了一种方法——探究物理/数字平台的关系及转化过程的方法，将讨论推进到以物体为中心的层面，在此讨论中，人和机器的联系变得更为紧密。

泰斯特将机器人实验室平台概念转化为对物体的一种"图像诱导性机器"操作[13]，巴特莱纳和纽斯曼更喜欢将之称为一个工具或一个舞台。他们的兴趣在于激活人的角色，并将他/她直接地带入到场景中去。将人或机器人代理整合到项目中是和项目本身同等重要的事情，或者更进一步说，整合过程就是项目本身。

构思环境

设计师及其使用工具之间的关系在不断发生变化。由于我们设计的系统变得更加复杂，必须将设计师纳入考虑之中，因此设计出自本真的直观、有趣才不会迷失在纷繁复杂的信息和界面中。机器人运动控制与建筑设计师日常使用的Maya，Rhino等现有设计工具的结合，可减少学习过程中的弯路，并使得设计师掌握如何在设计过程中充分利用机器人这一工具。与其他进行机器人建造研究的研究机构及初创公司相似，设计过程结束及制作过程开始之间的界限变得越来越交错不分，通常可以通过定制软件开发进行调整。美国南加州建筑学院提出的概念设计手法与软件开发密不可分。

Esperant.0 项目

项目始于2011年初，使用机器人实验室的第一组学生——克瑞姆·巴特莱纳，布莱登·克瑞斯曼和乔纳森·普罗托开发出一组Autodesk Maya插件，该插件弥补了工业机器人与设计师之间的距离。基于四维环境中关键帧控制特性，插件从基于文本的编程转化为可视化编程，使得设计师无需任何编程经验下即可在设计过程中指挥机器人运动。通过将机器人动力学与时间线

have been at the core of much of the work. While these approaches have proven to have explosive positive potential for the development of ideas as well for the development alternative skill set, they have also been subjects for discussion as parallel initiatives in the school vote for a more direct approach to the production of objects. The recurring question of scalability of robotic applications in this context is important as it pressures to have a critical assessment of the experiment.

A new generation of faculty and staff, Curime Batliner, Michael Jake Newsum and M. Casey Rehm, have recently entered the school and propose new angles in this discussion. Reinforced by observations in studios and seminars that projects with a critical amount of real world constraints as well a higher degree of interactivity in their processes seem to thrive, these educators promote an approach that interrogates the relationships and translation processes of the physical/digital platforms. They advance the discussion, which mostly revolves around objects, to a discussion where the relationships between humans and machines are becomes more relevant.

> 事实上，隔离模拟过程是不可能的；通过……反之亦然，……即不可能隔离现实过程或证明现实。
> It is practically impossible to isolate the process of simulation; through,…the inverse is also true,…,namely it is impossible to isolate the process of the real. or to prove the real. [11]
>
> ——让·波德里亚 Jean Baudrillard (1988)

Testa conceptualizes the Robothouse platform as an "image-inducing machine" [13] operating on objects, while Batliner and Newsum prefer to refer to it as a set or stage. They are interested in activating the role of the human and bringing him/her onto the scene more directly. The integration of human or robotic agency into the project is as important as the project itself or – to go further – it is the project.

Ideational Environments

The relationships of designers and their tools are continually evolving. As the systems that we design become more complex, the human designer must be considered, so that the intuitive and playful nature of design is not lost in the complexities of information and interfaces. The integration of robotic motion control into existing design tools that architects and designers use daily, such as Maya and Rhino, reduces the learning curve, and lets designers understand how to leverage the robot as a

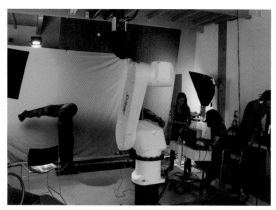

2011年，在为论文项目的同步对象进行移动位置的匹配
Match moving setup for thesis project synchronous objectives, 2011

我们面临的最紧急的技术问题就是在不采用已经开发利用的、不完整的人类系统的前提下，进行先进科技系统的人文网络建设，这是一个值得建筑科学理论家及梦想家关注的问题。

The most urgent technological problem facing us is the humane meshing of advanced scientific and technical systems without imperfect and exploited human systems, a problem worthy of the best attention of architecture's scientific ideologues and visionaries. [12]

——罗伯特·文丘里 Robert Venturi (1950)

结合（每1/24s），该插件使得设计师可同步协调及操作控制多个机器人，实现了机器人的复杂编排。结合一台摄像机和空间的精确测绘，该插件可匹配机器人实验室内的数字及物理信息，将机器人转换为精确的运动控制装置。Esperant.0项目杂交数字和物理工作流程，混合真实和虚拟内容，从而将整个机器人实验室空间转化为二元数字/物理设计界面。

Crane 项目

从2012年开始，布莱恩·哈姆斯在安德鲁·阿特伍兹，布莱登·克瑞斯曼和乔纳森·普罗托的支持下，继续研发适用于Rhino-Grasshopper的机器人运动控制插件。参数化建模插件为史陶比尔机器人手臂提供定制IK运算、模拟、测试信息，以及一键式的文件生成。Crane项目是对Esperant.0项目的回应，在性质上更加自由，但缺乏对大型脚本用户友好的程序化操作。它进一步发展了很多没有正式融入Maya插件的理念，如机器人流态位置控制以及实时监控。

tool in their design process. Similar to other research institutions and startups invested in robotic processes in fabrication, the limits of where design stops and production starts are becoming increasingly intertwined and are often mediated by custom software development. SCI-Arc's conceptual approach to design cannot be separated from the development of software.

Esperant.0

Starting in 2011, the first group of students to use the Robothouse, Curime Batliner, Brandon Kruysman and Jonathan Proto developed a plugin in for Autodesk Maya which bridges the gap between industrial robotics and designers. This shift away from text based programming and employing visual programming, through keyframing rigged characters in a four dimensional environment, lets designers incorporate robotic motion in their design process without any prior programming experience. By binding the robotic kinematics to a timeline every 24th of a second, the plugin gives the designer synchronous coordination and operational controls for multiple robots consequently allowing complex choreographies of the machines. In combination with a camera and the precise mapping of the space, this plugin transforms the robot into an excellent motion control device for match moving of digital and physical information within the Robothouse stage. Esperant.0 was the idea of hybridizing digital and physical workflows, mixing real and artificial content, thus turning the entire space of the Robothouse into dualistic digital/physical design interface.

Crane

Starting in 2012, Brian Harms continued to develop a robot motion control suite for Rhino-Grasshopper along with support from Andrew Atwood, Brandon Kruysman and Jonathan Proto. The parametric modeling plugin provides custom IK solving, simulation, diagnostic information and one-click file generation for Staubli articulated robot arms. Crane was a response to Esperant.0 which is more freeform in its nature but lacked the ability for large scripted user friendly procedural operations. It developed further many ideas that had not been formally integrated into the Maya plugin such as streaming position control over the robots as well as live monitoring.

Live

The streaming aspects of Crane were further developed in 2014 by SCI-Arc educators, Batliner and Newsum, into the platform "Live". Live situates the robot as an interactive design tool that immediately responds to designed inputs. Without a required preprogrammed motion or series of operations, the robot can engage directly with the current context available to the

2014年的"机器眼II"项目,人机交互的机器人驱动薄膜
Seminar Eyerobot II, robotic actuated membrane—human-robot interactions, 2014

programmed logic. It was developed to interact with any scripting language which opens the Robothouse, as a design platform, to designers of varying backgrounds and approaches.

The Live platform is a direct response to the way we designers work with robots in the Robothouse. Synchronous robotics concepts were pushed beyond robot to robot interactions as designers broke down barriers between robot operator and operation. In the traditional robot cell, robots work alone, void of human interference as operations are completed in repetitive sequences. This model does not hold validity in stochastic processes where the designer needs to make decisions on the fly which can modify the robot's immediate behavior. The programming for Live has alternative goals where the robot is no longer programmed to be optimized for repetition and precision. Instead, the robot is programmed to be versatile, nuanced and interactive.

Live 项目

2014年,美国南加州建筑学院的教师克瑞姆·巴特莱纳和迈克尔·杰克·纽斯曼对Crane项目的流程操作特性进行了进一步研究,将其融入"Live"平台。Live项目将机器人定位为交互设计工具,可对设计输入提供及时的响应。在没有预编动作或系列操作的情况下,机器人可以直接参与复合编程逻辑的当前环境。它可以与任何脚本语言进行交互,将机器人实验室设计平台向各种背景和专业的设计师开放。

"Live"平台是对我们设计师与机器人实验室的机器人一起工作的方式的直接响应。由于设计师打破了机器人操作员和操作之间的障碍,同步机器人技术已经超越了机器人,发展到了机器与人的互动阶段。在传统机器人室中,机器人通常单独工作,不会有人工干预,操作以重复顺序完成,这种模式对随机过程无效。在随机过程中,设计师需要在工作过程中做出决策,改变机器人的即时行为。Live编程提供了另一种目标,机器人不再在重复性和准确性方面进行优化;相反,机器人被设计为多功能的、精细的,以及交互式的。

为了使机器人沿着一条路径平滑的移动,机器人需要在机器人存储中设定不止一个位置点。运行Live时,机器人在线等待指令,机器人会根据接收到的顺序执行指令,并将其加入存储器。在存储中的指令被执行的时候,设计师可以随时干预或更改机器人操作。清除和重新填充存储器使设计人员能够直接控制机器人的操作,这使得机器人通过将应用程序的指令更新到当前的操作环境,不断地进行随机操作。

"Live"平台是"实时、仿真、模拟和后动画"的突破点,或者说"Live"平台瓦解了"实时、仿真、模拟和后动画"[16]。由于机器人可以在响应实时指令的过程中修改其预设的操作,机器人运行不再是确定性的或可模拟的。因为模拟不能预测影响机器人行动的环境条件,所以机器人操作模拟不能再对实际的机器人程序进行干涉。后动画变得更加有效,它不再是对模

居住空间超越了几何空间……如果说房子是孩子们最先接触的世界,那么空间形状又是如何影响人们对于空间和更大世界的认识的呢?房子是"有机行为组",还是有着更为深刻的内涵,是想象力的居所?

Inhabited space transcends geometrical space... If the house is the first universe for its young children, the first cosmos, how does space shape all subsequent knowledge of other space, of any larger cosmos. Is that house "a group of organic habits" or even something deeper, the shelter of the imagination itself? [14]

——约翰·斯蒂尔格 John Stilgoe (1994)

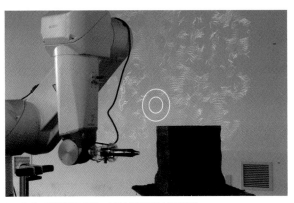

2014年的"机器眼II"项目,人机交互的机器人驱动薄膜
Seminar Automaton, Group 1, live agency-driven fabrication, 2014

拟进行仿真,而是集中于事件物理状态的一体化呈现。实体机器人不再是数字设计操作的模拟,相反,实体机器人和实时操作可以作为重要的设计工具。

"机器眼 II" 项目

由巴特莱纳授课的2014春季"机器人眼 II"课程,以及由莱姆授课的2014秋季"自动化:设计智能"课程,都是运用了实时定位控制的范例。课程中开发的软件发展成为"Live"机器人控制平台。

"机器眼II"项目的最终展品为一个原型装置,探索人在带有可变动的响应式表皮的空间中的行为,同时探究空间的内在逻辑。展览位于机器人实验室,参观者受邀进入受监控的夹道空间,这个空间由现有室内墙体和中央一个临时构筑物界定。室内构筑物被一层薄膜覆盖,薄膜由能够改变薄膜几何形状的动态支架支撑。支架本身为四个工业机器人,以松散的形式与薄膜连接。机器人移动状态与可感应人类活动的内置即时感应器相关联。展现在薄膜上的即时一系列视觉画面有助于提高装置的响应质量。

在无人在场的情况下,机器人按照预设程序运行,通过处理薄膜来改变空间。即时视觉画面可延迟响应,展现表面运动的历史记录。一旦有人进入该空间,机器人即转入即时模式,跟踪人体移动。此时,空间为激活状态,围护结构造型、即时视觉画面都与人们活动相关联。

进入本空间的绝大多数参观者都不了解规则设定。部分参观者到处摸索或寻求说明解释直至其适应展品活动状态。一旦人们感到其可对此进行控制,他们即从起初较被动的行为转变为轻松愉悦的互动。部分参观者只是来回走动,并观看其行为结果,但是绝大多数参观者改变自己的行为使得机器按照其预想的路径进行运动,并最终改变空间质量。

In order to produce a smooth motion along a path the robot needs more than one position in the robot's memory. While running Live, the robot is online waiting for commands that will be executed according to the sequence in which they are received and added to the memory. As the commands in the memory are executed, the designer is able to intervene or alter the robotic operations at any time. Clearing and refilling the memory gives the designer immediate control over the robot's operations which allows the robot to continually engage in stochastic processes by updating the application's commands to the current context of the robotic operation.

The Live platform is a breaking point from or a collapse of the "Real-Time, Simulation, Emulation and Post Animation" [16]. The robot's operations are no longer deterministic or simulatable

> 机器人制造的建设性能力清晰界定了设计空间,并颇有成效地促使计算设计策略的提出,机器人的开发可被公认为创造性设计行为。
>
> As the constructive capacities of the robotic fabrication process clearly define the design space and thus productively inform the computational design strategy, its development can also be recognized as a creative act of design on its own. [15]
>
> ——迈克尔·布迪 等 Michael Budig et al. (2014)

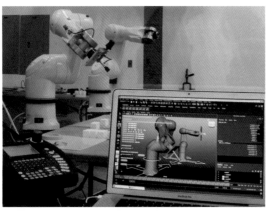

在2011年的Esperant.0项目中,一款Maya的插件解决了工业机器人和设计师之间的互通问题
Esperant.0, Maya-Plugin which bridges the gap between industrial robotics and designers, 2011

since the robot can modify its predetermined operations as it is responding to live instructions. While the emulation of robotic operations can no longer anticipate the real robotic procedures that will occur because of the emulation's inability to anticipate the contextual conditions that will inform the robot's actions, the post-animation gains validity as it moves beyond a role of calibrating the simulation into the role of digital integration of the physical stage of events. No longer is the physical robot an emulation of the digitally designed operations, but instead the physical robot and the live operations can be engaged as the primary design tool.

Eyerobot II

"Eyerobot II", a seminar taught by Batliner in spring 2014, and "Automaton: Designing Intelligence", taught by Rehm in fall 2014, are case studies that utilize the real time position control. The software developed throughout these seminars, has since evolved into the robot live control platform "Live".

"自动"项目

另一个课程"自动:设计智能化"旨在研究将智能代理运用于机器人控制平台,探讨了即时反应系统的内在潜力和问题。所有的项目通过把一系列分析和决策功能转化为机器人动作的控制数据,利用Processing中开发的控制平台来链接规则(本项目中指Microsoft Kinect及机器人在现实世界中的位置)。学员们并非设计预先设定的结果,再开发机器人路径以达到目的;而是利用基于智能代理研究的模型来设计机器人的反应行为,以便对不断变化的环境做出反应。这种练习不再是设计具体物体然后去制作;相反,设计者们将他们的意图和形式构想以编码的方式写入制作装置中,练习的目标是将设计者的决策延伸到能够解释、结合材料,并超越人类能力的装置中,为建筑师和设计师提供了另一种操作模式。

文化适应

"机器人如何通过增加材料的差异性及形式的复杂性,为建筑师扩大生产及设计范围?" [17] 这个问题以及机器人如何直接影响建成环境的问题,体现了建筑机器人话题的外在特性。在这个背景下,机器人实验室的方法论重点关注当代设计师的发展,设计师具备特定技能,不仅可以影响建成环境,还可以参与到其他形式的文化生产中。

机器人在当前对于实体及文化产品的探讨中发挥着重要作用,且具有出色的教与学能力。但随着技术的发展及变化,机器人可能会消失。

The final exhibit of Eye Robot II was a prototype installation which started to explore the behavior of humans in spaces with transforming, responding envelopes as well questioning the underlying logic of such spaces. The exhibit was located in the Robothouse. Visitors were invited to move through a monitored in-between space which was framed by the interior walls of the existing space and a temporary volume in the center. The interior volume was surfaced by a thin membrane that was held by a dynamic scaffold which allowed for a transformation of the surfaces' geometry. The scaffold itself was four industrial robots with loose connections to the membrane. The movements of the robots were linked to live sensor inputs which respond to human activity. A set of live visuals rendered onto the membrane amplified the responsive quality of the installation.

With no human present, the robots were running on a preprogrammed routine deforming the space by manipulating the membrane. The live visuals responded with a delay and displayed the history of the surfaces' motion. Once a human entered the space, the robots switched into live mode, tracking the human's body. Now the space was active and the shape of the envelope and the live visuals linked to the human movement.

Most visitors entered the space with no instructions and knowledge of the rule set. It took some feeling around, or explanation, until people started to feel comfortable with the active state of the exhibit. Once people felt in control, their initial, more

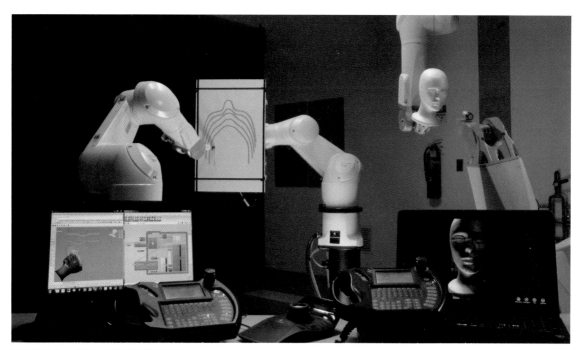

在2014年的Crane项目中,对数字空间进行了转译
Crane, digital-spacial translation, 2014

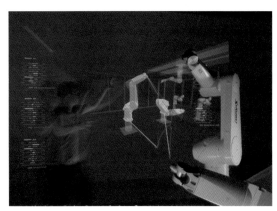

在2014年的"现场"项目中,运用VVVV控制机器人基于手势的运动
Live in action, gesture-based robot motion control using VVVV, 2014

对建筑的野心比建筑技术更有价值。
An architectural ambition precedes an architectural technology. [18]

——杰利·法瑞加 Jelle Feringa (2014)

从受包豪斯车间影响的课程中汲取的经验及知识已经超越了技术本身,比研究机器人的复杂性更进一步。作为数字/物理界面,机器人实验室平台总结出一套设计流程的整体方法,促进了团队协作、跨专业设计开发,解决了现实世界中的局限,清除了沟通障碍。

显然,建筑师的方法及专业正在以令人兴奋的方法进行重塑。"设计生产领域多个部分的协同工作,……旨在打造出'高度信息化'的建筑,以及高度信息化的'建筑师'。" [19]通过机器人,企业家精神也蔓延至了建筑领域。许多机构(如苏黎世高工的建造实验室和斯图加特大学的计算设计学院)正掀起新创业公司的潮流,目标直指建筑领域;南加州建筑学院机器人实验室的大量毕业生也创办了自己的技术公司——Thesugarlab(现在属于3D系统的一部分)及ID4A,或者在创新型公司中担任重要角色,如Bot&Dolly事务所、三星及耐克等,虽然这些公司在建筑空间生产中可能并未发挥主要角色,但有人认为这些企业对于社会文化甚或对于社会的影响十分重要。

passive, behaviors changed into a playful interaction. While some visitors just moved around and looked at the consequences of their actions, most visitors altered their behavior to make the machine do what they wanted it to do, ultimately transforming the quality of the space.

Automaton

Another seminar, Automaton: Designing Intelligence, looks at utilizing intelligent agency within the robot control platform to explore the potentials and issues inherent in live responsive systems. All projects leveraged a control platform developed in Processing to link precepts, in this case a Microsoft Kinect and the robot's real world position, through a series of analytical and decision making functions which are then translated into robot motion control data. Instead of designing an intended outcome and then developing robot motion paths to achieve that goal, students designed the behavior of the robot's response utilizing models based on research of intelligent agents to engage constantly changing contexts. This exercise becomes less about designing a specific object and then fabricating it. Rather, the designer codifies his/her intentions and formal intelligence into the fabrication device, the goal being to extend a designer's decisions into a device which can interpret and engage materials and contexts with non-human capabilities. The project collapses simulation, representation and fabrication within the design process, offering an alternative mode of operation for architects and designers.

Enculturation

"How can robotics expand the range of production and design options for architects by increasing the potential of greater material differentiation and complexity of form?" [17] This question, along with questions of how robots will directly influence the built environment, represent the outward identity for the discourse of robotics in architecture. Under the hood, the Robothouse's methodologies are focused on the development of contemporary designers that are equipped with the right skill sets which allow them not just to influence the constructed environment but also to participate in other forms of cultural production.

Robots are playing an important role in the current discussions about physical and cultural production, as well as serving as an excellent platform for learning and teaching, but as technology evolves and changes, robots might disappear.

The lessons learned and the knowledge developed in these Bauhaus-workshop-inspired courses transcends the technology and go well beyond learning the intricacies of the robot. The Robothouse platform – as a digital/physical interface – nurtures a holistic approach to design processes, which promotes teamwork, interdisciplinary idea development, working with real world constraints and clear communication.

It is apparent that the approach and professional outlook of architects are being reshaped in exciting ways. "Simultaneous occupation of multiple phases of the design-production spectrum,…, seeks to create not only 'highly informed' architecture, but highly informed 'architects'" [19]. Entrepreneurship finds its way through robotics into the architectural field. While institutions such as the FabLab at ETH in Zurich or ICD in Stuttgart create a constant stream of new startup companies targeting directly the building industry, the Robothouse has produces alumni who have started their own tech-startups; thesugarlab (now part of 3D Systems) and ID4A or taken roles in innovative companies; Bot & Dolly, Samsung and Nike. Although these companies might not play a significant role in the production of built space, one might argue that their reach into our social culture and therefore on society are equally important.

参考文献 / References：

[1] Gramazio, Fabio, and Matthias Koller. "How Robots Change Architecture." The Robotic Touch (2014) 10.
Gramazio, Fabio, Matthias Koller, and Jan Willmann. "Introduction." Architectural Design 229 (2014): 14-23.

[2] Khun, Thomas. "Scientific Revolutions." Philosophy of Science. (1991): 75. Referenced in Verebes, Tom. "Counterpoint." Architectural Design 229 (2014): 126-133.

[3] [5] [8] [13] [16] Testa, Peter. "Autonomous Translations." Fabrication and Fabrication. SCI-Arc Press (2014): 42-51.

[4] [7] [9] Sloterdijk, Peter. "Spharen." Globen, Vol. 2 (1999). Quoted in Marc Jongen's, "On Anthropospheres and Aphrogrammes. Peter Sloterdijk's Thought Images of the Monstrous." Humana Mente Journal of Philosophical Studies, Vol. 18 (2011): 199-219.

[6] Wolf, Amit. "Introduction." Fabrication and Fabrication. SCI-Arc Press (2014): 6.

[10] Tschumi, Bernard. "Responding to the Question of Complexity." Complexity. Art, Architecture, Philosophy. Journal of Philosophy and the Visual Arts. No. 6 (1995) 82. Quoted in Gleininger, Andrea, and Georg Vrachliotis. "Editorial." Complexity Design Strategy and World View (2008) 7-11.

[11] Baudrillard, Jean. "Simulacra and Simulations." Selected Writings / Edited and Introduced by Mark Poster. (1988): 179.

[12] Venturi, Robert. "Context in Architectural Composition: Excerpts from M.F.A. Thesis, Princeton University, 1950." Complexity Design Strategy and World View (2008) 13-23.

[14] Stilgoe, John. The Poetics of Space: The Classic Look at How We Experience Intimate Places. Beacon Press, 1994.

[15] Budig, Michael, Jason Lim, and Raffael Petrovic. "Integrating Robotic Fabrication in the Design Process." Architectural Design 229 (2014): 23.

[17] Castle, Helen. "Editorial." Architectural Design 229 (2014): 5.

[18] Feringa, Jelle. "Entrepreneurship in Architectural Robotics: The Simultaneity of craft." Architectural Design 229 (2014): 60-65.

[19] Johns, Ryan Luke. "Greyshed." Architectural Design 229 (2014): 74-75.

2013-14 ICD/ITKE研究展馆
ICD/ITKE Research Pavilion 2013-14

Explorative Design and Fabrication Strategies for Fibrous Morphologies in Architecture
建筑纤维形态学的探索性设计与建造策略

Moritz Dörstelmann, Marshall Prado, Achim Menges / ICD Institute for Computational Design, University of Stuttgart
莫瑞斯·多斯特曼，马歇尔·普拉多，阿希姆·门格斯 / 德国斯图加特大学计算设计学院

生物体在漫长的进化过程中发展出了其结构原则，这些结构原则在个体发育过程中，演变成了适应当地语境的生物形态；由此所生成的形态综合了大量内在和外在的准则。资源可用性作为重要的边界条件之一，高效能材料的开发对一个物种的整体性能而言至关重要。在此过程中，性能化形态的开放式发展及材料组织构成了两大主要因素。总的来说，对材料属性的局部改变，是提升资源利用率的主要驱动力之一。而各向异性材料的差异化组织能够使材料最大限度地适应结构和功能的需求。毫不奇怪，最终，随着时间的推移，大部分生物进化而来的结构都包含纤维复合材料。胶原纤维具备广泛的材料特性，既有可变的结缔组织，也可通过纤维方向的变化，在肌腱处形成定向的高强度抗拉筋。植物通常根据压力或张力荷载调整纤维素纤维的方向；节肢动物通过及甲壳素纤维的差异化组织形成一系列适应当地的壳体结构。

在技术运用中，纤维增强复合材料（FRP）充分利用了各向异性材料卓越的结构性能优势。从二十世纪三四十年代早期开始，人们就已对纤维增强复合材料在建筑和工程中的应用进行了实验测试 [1]。在自然的形态生成过程中，材料的各向异性特质通过局部差异化可以发挥最大的潜力；但那时，技术复合材料缺乏被设计和建造成复杂形式或纤维组合的方法。而如今，计算设计、墨迹工具以及先进的数字建造方法使得纤维增强复合材料可以通过经济可行的方式实现局部差异化。计算设计策略使得性能化几何体的开发成为可能，同时将各种设计驱动因素整合进纤维增强复合材料之中。通过与创新的数字建造策略相结合，这些先进技术及其在建筑设计领域的应用为自下而上

Biological evolution allows structural principles to develop over long periods which then unfold into locally adapted and contextualized solutions within an ontogenetic process. The resulting morphologies synthesize a multitude of intrinsic and extrinsic criteria. With one of the critical boundary conditions being the availability of resources, the development of material efficient solutions becomes critical for the overall performance of a species. Two main factors in this process are the open ended development of performative morphology and material organization. While local variation of material properties, in general, is one of the main drivers for resource efficiency, differentiated arrangements of material with anisotropic characteristics, in particular, allows maximum adaptation to structural and functional requirements on the material scale. Ultimately it is no surprise that over time most biological organism evolved structures consist of fiber composite material. Collagen fibers can achieve a wide range of material behaviors ranging from flexible connective tissues to highly directional tension reinforcements in tendons through variation in their fiber directionality. Plants locally adapt the orientation of cellulose fibers in response to compression or tension loads and arthropods achieve a wide variety of locally adapted shell structures through differentiated arrangement of chitin fibers.

In technical applications fiber reinforced polymers (FRP) take most advantage of extraordinary structural properties of anisotropic materials. FRPs have been experimentally tested

地研究纤维材料系统和建造策略打开了巨大的可能性。由此，高度灵活的纤维复合材料构件的生成和建造成为可能又进一步推动了节约资源的建造原则及功能一体化的概念从生物先例到建筑应用的仿生学转化。这些过程不仅能提高材料利用率、促进功能一体化；而且在创新的建构学中体现出性能化原则及建造过程，形象地从建筑细节和表皮阐述中体现出材料特性。这些建筑成果最终开拓出了一片建筑设计行业的新天地。

高承载力及易成形性使得纤维增强复合材料非常适用于工程应用。制造这些复合材料的传统方法，包括在全尺寸表面模具上层压树脂基体中的纤维材料。这对性能化建筑设计带来了挑战：复合材料的结构条件要求具备可控的纤维排布以及几何差异形式的定制。而纤维增强复合材料适用于塑造各种形状，是实现这种目的的理想材料，然而考虑到经济问题，制造一个适用于各种几何形式的模具，对大规模生产建筑尺度的复合材料来说并不可行。这些工艺通常受限于批量生产出相同的零件以便更好地利用制造模具所需的初始资金和材料。具备不同纤维排布的复合材料的生产同样也是一种劳动密集型的工艺，很难进行手工操作。当然，机器人建造工艺能够精确地排布复合材料结构中的纤维，这使得自定义建造成为可能。

在一些建造先进复合材料结构的行业中，如航空、造船和汽车业，已有一些运用数字化建造技术进行纤维铺放的实例。流程采用了由机器人控制芯轴的大型编织机，斯图加特大学飞机设计研究所（IFB）采用这样的机器，因为这种机器能够塑造复杂的几何形体，而不会导致织物的悬垂或起皱 [2]。在此，机械臂能够在编织过程中控制芯轴的速度和方向，从而控制纤维的方向和密实度。纤维围绕闭合模架不断缠绕，不会产生影响结构承载力的接缝。纤维变角度牵引铺放技术（TFP）是另一种数字化建造工艺，可以将增强纤维缝制在织物底板之上，使纤维的方向和密度具有高度的灵活性 [3]。然后将这些强化后的部分用树脂浸染，覆于某模具之上形成复杂的几何形体，可以经过铣削后期加工来实现 [4]。同传统层压技术一样，这些工艺仍然需要纤维模具来形成复杂形态，当然也可通过铣削加工而成，但有可能影响纤维的完整性。虽然3D打印技术无需任何模具便可生成复杂的几何体，但是直到最近，打印复合材料仍旧仅限于短纤维的使用，与连续纤维复合材料相比，短纤维降低了材料的结构性能。MarkForg3D公司声称其已经制造出世界上第一台能够打印连续碳纤维的3D打印机(markforged.com)。目前该技术仅用于面层方向上具有连续纤维的分层熔融沉积组件。TFP技术和3D打印复合材料技术对纤维方向都具有敏锐但却有限的控制，且二者能够制造的连续纤维组件的尺寸也都有限。数字化建造的复合材料类型多样，能用于多种性能化领域，但受到尺寸限制，需要模具或后期加工才能完成，因此这些技术似乎都不适合性能化建筑应用。

为了减少建筑尺度的复合增强纤维材料结构在建造过程中所需的不必要的模具，斯图加特大学计算机设计学院（ICD）及结构建造和设计学院（ITKE）共同研发出了一套纤维铺放的创新流程。空心纤维缠绕技术是一种纤维铺放的系统流程，无需表面模具。斯图加特大学2012 ICD/ITKE研究展馆，首次将空心

for architectural and engineering applications since the early 30s and 40s [1]. But unlike natural processes of morphogenesis, where the anisotropic character of the material can be used to its highest potential through local differentiation, technical composites were, at the time, lacking methods to be designed and fabricated in complex forms or fiber arrangements. Today computational design and simulation tools as well as advanced digital fabrication methods allow differentiation in FRP parts in an economically feasible way. Computational design strategies enable the development of performative geometries and the integration of various design drivers into the material layup of FRPs. In combination with novel digital fabrication strategies these advances and their application in architectural design open up a large potential for bottom up explorations of fibrous material systems and fabrication strategies. The resulting possibilities for the generation and fabrication of highly articulated fiber composite parts open up the potential for biomimetic transfer of resource efficient construction principles and functional integration concepts from biological precedent into architectural applications. These processes not only facilitate material efficiency and functional integration, but manifest performative criteria and fabrication processes in novel architectural tectonics and visually express material characteristics in architectural detail and surface articulation. These architectural implications ultimately allow us to explore and expand a new field within the architectural design repertoire.

High load bearing capacity and ease of formability make FRPs well suited for engineering based applications. Traditional means of fabricating these composites include laminating fibrous materials in a resin matrix on a full scale surface mold. This poses a challenge for performance based architectural design where structural conditions require controlled fiber arrangements and tailoring of geometrically differentiated forms. As FRPs are well suited to take a wide variety of shapes, they are an optimal material for this purpose. However, manufacturing a mold for all variations in geometric form is not feasible for large scale production of architectural scale composites without financial tradeoffs. These processes often restrict the outcome to a series of mass produced identical parts in order to take advantage of the initial financial and material investment required to produce such molds. The production of composites with differentiated fiber arrangements is also a labor intensive process that is difficult to implement manually. There are, of course, robotic fabrication processes for the precise arrangement of fibers in composite structures that make custom fabrication possible.

Several examples of digitally fabricated fiber placement are being utilized for the fabrication of advanced composite structures in industries such as aerospace, naval architecture and automotive. These processes include large scale braiding machines with

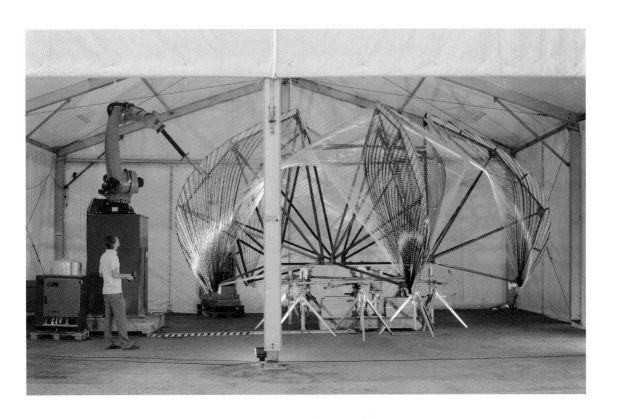

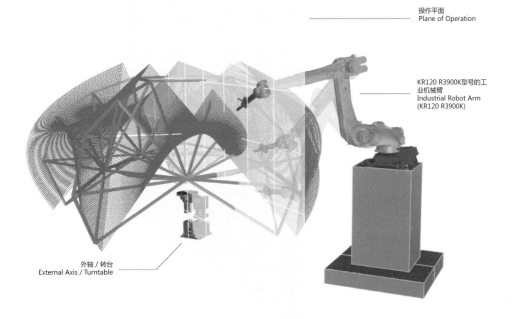

操作平面
Plane of Operation

KR120 R3900K型号的工业机械臂
Industrial Robot Arm (KR120 R3900K)

外轴 / 转台
External Axis / Turntable

2012 ICD/ITKE研究展馆：通过交互式模拟来实现机器人建造的7轴运动轨迹，并完成其现场的调试和代码的生成。全尺寸的机器人缠绕在旋转的临时框架上实现。
ICD/ITKE Research Pavilion 2012：Interactive simulation of the 7-axis kinematics of the robotic fabrication setup and code generation. Full-scale robotic filament winding onto the revolving temporary framework.

纤维缠绕技术应用于大尺度的纤维复合材料建筑中。该流程采用机器人设置，将树脂浸泡的纤维缠绕在最小化的大型旋转框架上[5]。复合材料结构的几何曲面不再依赖于所使用的模具，而是受制于随后铺放并受压的纤维之间的相互作用。形成复合表皮的缠绕句法或系统特定于卷绕架几何的缠绕顺序。采用这一方法建造的结构规模仅受限于机器人设置。2012 ICD/ITKE研究展馆为8m的单体式复合壳体，利用六轴机械臂和旋转定位器。生成这种性能化复合曲面需要几何卷绕架、纤维缠绕顺序和系统张力的协同。通过适当协调各成分变量之间的关系，纤维之间能产生足够的相互作用以形成稳定的复合结构，并通过均衡材料应力分布自发形成复杂的双曲面。

曲面几何和纤维方向丰富多变，为了设计丰富的相互依赖关系并协调可能发生冲突的设计变量、目标和限制因素，一种综合性的计算设计方法被开发出来[6]。针对这一流程研发的数字工具能够在有限的设计空间中迭代并探索。建造限制、结构性能、材料特性以及生物设计原则同时成为设计驱动因素，通过协调因素间复杂的相互关系，生成性能化几何体。

该综合流程的研发融入了多个学科的知识，并突出交叉学科所具有的创新潜力。跨专业团队的每个成员不应仅仅具备各自特定的专业知识，而且应该具有跨学科的认识和概念抽象的能力，从而理解相互间的沟通。对问题、方法和技术的抽象有助于识别概念和技术转化的潜力区域，为跨学科转化、改造设计方法带来启发；或是已有技术的新应用或是某干先前背景，改造技术以实现新用途。例如，纤维增强复合材料结构最初是由航空航天业转向建筑业的，在建筑语境中对这一结构的建造策略和技术的自下而上的探索，带来了航空航天业的相关发展，引起生成知识的逆向转化。在仿生学研究中，也遵循类似从调研、抽象概括、转化和逆向转化的跨学科流程。对生物先例的调研激发对其潜在功能原则的抽象概括，进而将功能原则转化为技术应用。在该流程中，用于生物学调研的方法或工具也能透过多学科的镜头，生成有价值的生物理想模型。

这些方法有助于进行物质化策略及生物轻质结构原理之间自下而上的相互调研。鉴于生物概念的复杂程度通常超过了现有技术所能实现的差异化水平，对材料的新用途及操作流程的探索能够扩大生物概念转化的可行性。在这些展馆案例中，机器人建造策略和纤维复合材料的结合以及它们向建筑领域的转化造就了新的结构差异化程度，并因此开拓了在实际工程中应用生物原则的潜力。

在2013-14年的ICD/ITKE研究展馆项目中，通过对空心纤维缠绕方法及生物纤维复合结构的研究，使得由生物学家、古生物学家、建筑师和工程师组成的跨专业团队研发出了一个模块化轻量级纤维复合建筑系统，并通过对全尺寸原型的测试，证明该体系可用于建筑领域中。

对天然纤维复合结构的功能原理研究，证实了飞行甲虫物种的后部保护翼壳（翅鞘）为生物纤维复合结构中轻质结构的理想模型。不会飞的甲虫物种具有实心的复合壳体，这种外壳仅能容纳

2012 ICD/ITKE研究展馆
ICD/ITKE Research Pavilion 2012

robotically controlled winding mandrels,. The Institute for Aircraft Design (IFB) at the University of Stuttgart utilizes such a machine due to its capacity to cover geometrically complex forms without the draping or wrinkling of fabric [2]. A robotic arm is utilized in this case to control the speed and orientation of the mandrel throughout the braiding process, thereby controlling the fiber orientation and compactness. Fibers are wrapped continuously around closed forms without the need for structurally inefficient seams. Tailored Fiber Placement (TFP) is another digital fabrication process in which reinforcement fibers can be stitched onto substrate fabric providing a high degree of flexibility for fiber directions and densities [3]. These reinforced parts can then be resin impregnated over a form or post processed through milling into complex geometries [4]. These processes, like traditional lamination, still require a fiber mold for complex forms or otherwise through the milling process, run the risk of compromising the integrity of the fibers. Three dimensional printing, however, is a technology which does not rely on molds to create complex geometries. Until recently printing composite materials was limited to using short fibers which reduces the structural capacity of the material compared to continuous fiber composites. MarkForg3D is a company that claims to have created the world's first continuous carbon fiber 3D printer (markforged.com). This technology is currently limited to layered fused deposition parts which have continuous fibers in the orientation of the layers. Both TFP and 3D printed composites have acute, though limited, control of fiber orientation and are constrained in size for components with continuous fibers. The wide variety of digitally fabricated composites are suitable for many performative applications but due to the limitations in size, mold requirements or necessity for post processing none of these options seem suitable for performative architectural applications.

The Institute for Computational Design (ICD) and the Institute of Building Structures and Structural Design (ITKE) at the University of Stuttgart have developed an innovative process for fiber

薄管液体传输（血液淋巴）和气体传输（呼吸系统）。不同于这些陆生物种，飞行甲虫不仅需要维持其壳体的结构性能以便保护自己，同时还将飞行所需的能量压缩至最小以便更轻盈地飞行。这些看似矛盾的体质标准，在飞行甲虫翅鞘形态的进化过程中得到了协调，即将内部运输管道适应性变化为大空腔。翅鞘内外层之间的联系被简化为如支撑结构（骨小梁）一般的间柱。翅鞘形态的变化及骨小梁的分布与整体结构体系中的局部结构需求有关。借助扫描电子显微镜和微型计算机断层扫描技术，团队对多种飞行甲虫物种进行了比较研究。通过比较研究，得出了一系列抽象化形态及材料组织原则。基于这些原则，团队研发出模块化纤维复合结构系统，这一系统通过计算设计工具同时整合了机器人建造、材料特性以及结构因素。

不同于2012 ICD/ITKE研究展馆（通过定制建造设置来打造单一的纤维复合壳体），差异化组件的建造需要灵活的机器人设置以适应各种几何形态和特定的缠绕句法。团队引入两台机器人协作的空心纤维缠绕系统[7]。在此系统中，每台机器人都装配有2m直径的钢桁架效应器，为各种卷绕架提供用以附着的轻量稳定结构，这些效应器及卷绕架都必须有足够的刚性来抵抗缠绕过程中张拉纤维所产生的张力。这两台机器人必须同步运作以保持卷绕架之间的距离和方向，避免因缺少系统张力导致纤维变形或施加过多张力导致纤维断裂，从而破坏发展中的复合几何体。为了保持两台机器人的同步运作，其中一台被设计为主机器人，发布缠绕过程中的运动指令；而另一台为辅助机器人，跟随主机器人位置和方向的改变做出相应变化。这

placement in order to reduce the need for superfluous formwork in the fabrication of architectural scale FRP structures. Coreless filament winding, a methodical process for fiber placement which requires no surface formwork, was first developed for large scale architectural applications of fiber composites in the ICD/ITKE Research Pavilion 2012 at the University of Stuttgart. This process employs a robotic setup to wind a resin impregnated fiber around a large yet minimized rotating frame [5]. The surface geometry of the composites structure is no longer determined by the mold in which it is applied but rather the interaction of subsequently laid and tensioned fibers. The winding syntax, or the systematic sequence of wrapped fibers to form a single composite surface, is specific to the frame geometry it is wrapped on. The sizes of the fabricated structures utilizing this method are limited only by the robotic setup employed. The ICD/ITKE Research Pavilion 2012 was an 8 m monocoque composite shell that utilized a six axis robotic arm and rotational positioner. Development of the performative composite surface requires the negotiation of the wrapping frame geometry, sequence of wrapped fibers and the tension on the system. Proper coordination of the constituent variables can create enough fiber to fiber interaction to form a stable composite structure and generate, through the equalization of the material stresses self-forming complex hyperbolic surfaces.

A wide variety of surface geometries and fiber orientations

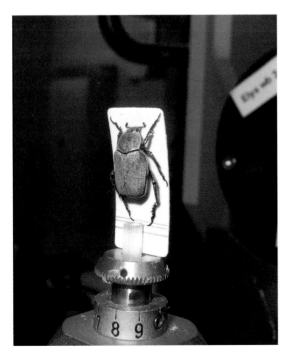

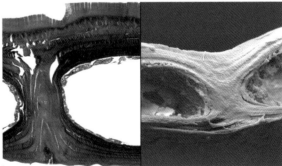

2013-14 ICD/ITKE研究展馆：微观计算机对飞行甲虫进行了断层扫描，以分析其纤维布局与结构形态的关系
ICD/ITKE Research Pavilion 2013-14 : Micro-computed tomography of the flying beetles elytron scanned for the project. Correlation of fiber layout and structural morphology in trabeculae. © Dr. Thomas van de Kamp, Prof. Dr. Hartmut Greven | Prof. Oliver Betz, Anne Buhl, University of Tübingen

两台机器人同步运作时可创造12轴动态系统。

为了进一步减小无芯缠绕系统所需的模板,采用小型可重复使用部件制作缠绕框架。每一组件在几何上由四和七顶点之间的两个非平面多边形确定。卡紧机械具有多自由度,以调节每一顶点平面的X、Y、Z坐标位及A、B、C定位。充分利用机器人的精密度,开发出准确操纵机器人的装配程序,从而使每一顶点夹具都能精确地附着于底座效应器上。一旦夹具准备就绪,5轴铣削木质参考线将在相邻多边形边缘上确定环绕点的位置和旋转角度。我们在这些参考线中置入尽可能多的信息以降低可调节卷绕架的整体机械复杂度。整合机器人装配式流程和可重构元件包意味着我们能够打造精准而坚固的卷绕架,且无需进行附加的测量或数字化操作。

装配式几何框架以及同步的机器人定位和动作使得差异化复合构件的缠绕成为可能。在两台机器人之间是固定的纤维源。树脂浸泡系统用以浸染玻璃或碳纤维粗砂[8]。两台同步的机器人将卷绕架旋转到位,以便纤维源顺次绕过每一个缠绕控制点。起初,纤维以直线的形式交替地从一个框架连接到另一个。随着缠绕在旋转框架上的纤维越来越多,纤维开始交叉层叠。最终,纤维之间的相互作用形成了一个复杂的双曲面。增加的多层玻璃纤维和碳纤维分别生成了构件的整体形态和特定的结构加固。每组缠绕代码、结构加固层、构件几何和连接节点在整体系统中都和相邻构件相协调。

每一构件的设计与建造都需要一套综合计算设计方法(ICDM)。

are possible. In order to choreograph the multifarious interdependencies and negotiate the variety of potentially conflicting design variables, objectives and constraints, an integrative computational design methodology was developed [6]. Digital tools developed for this process allow for the iteration and exploration of the constrained design space. Fabrication constraints, structural capacities, material behaviors and biological design principles can all be simultaneously considers as design drivers and through negotiation of the complex set of interrelations can a performative geometry be generated.

The development of such multifaceted processes involves the expertise of multiple professions and highlights the exemplary innovation potential situated at the intersections of various disciplines. A multidisciplinary team should not only include the specific expertise of each participant, but a cross disciplinary understanding and the ability for conceptual abstraction to establish a comprehensible level of communication. The abstraction of questions, methods and technologies allows for the identification of potential areas for concept and technology transfer. The adaptation of methods for cross disciplinary transfers can potentially generate insights in both disciplines including new applications for an existing technology or the adaptation of technology from a previous context for new uses. In the case of fabrication strategies and technologies for fiber reinforced polymer structures, which were initially adopted for architectural

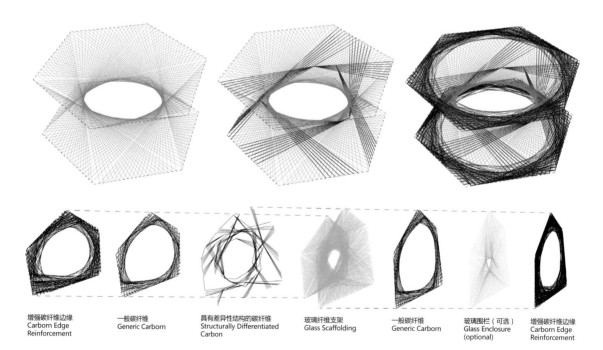

增强碳纤维边缘 Carborn Edge Reinforcement | 一般碳纤维 Generic Carborn | 具有差异性结构的碳纤维 Structurally Differentiated Carbon | 玻璃纤维支架 Glass Scaffolding | 一般碳纤维 Generic Carborn | 玻璃围栏(可选) Glass Enclosure (optional) | 增强碳纤维边缘 Carborn Edge Reinforcement

2013-14 ICD/ITKE研究展馆:一个单体中的纤维布局
ICD/ITKE Research Pavilion 2013-14 : Fiber layout for one component

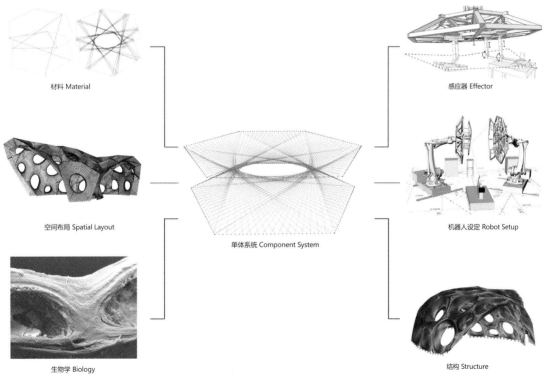

材料 Material

感应器 Effector

空间布局 Spatial Layout

机器人设定 Robot Setup

单体系统 Component System

生物学 Biology

结构 Structure

2013-14 ICD/ITKE研究展馆：基于单元组件的建造系统是多个过程参数的集成
ICD/ITKE Research Pavilion 2013-14 : Integration of multiple process parameters into a component based construction system

2013-14 ICD/ITKE研究展馆：双曲面的玻璃纤维几何和碳纤维的增强
ICD/ITKE Research Pavilion 2013-14 : Doubly curved glass fiber geometry and carbon fiber reinforcement

在2013-14 ICD/ITKE研究展馆中，这一方法通过多种途径来实现：首先，综合计算设计方法是集所有设计限制、目标和变量的生成性总体设计工具。我们采用了物理模型系统，将设计参数抽象为质点弹簧系统中的物理作用力。物理模型系统用于控制总体展馆中单体构件的几何形态。作用于任一构件的力也将同样作用于其周边构件。迭代的几何解决方案可由作用力之间总体平衡的状态来表示。其次是体现在模拟上。通过模拟机器人建造流程来测试12轴机器人装置的运动解决方案是否可行。过大的或是超出机器人装置动作范围外的构件几何体将通过生成性总体设计工具进行调整。设计流程中同样整合了结构性能模拟，以评估和调整每一次迭代的结构性能。最终，团队利用分析流程，通过性能化设计原则的筛选，生成和评估设计迭代。

因此，我们设计并建造了一座全尺度的演示展馆，它由36种几何形态各异构件组成，其中最大的构件直径约2.6m，重24.1kg。每一构件的缠绕都无需预制模板或昂贵模具。通过机器人建造策略流程创造出响应多样化设计和结构因素的不同纤维排列方式。最终的演示展馆占地50m²，体积约122m³。展馆非常轻质，重量总共不到600kg。

斯图加特大学计算设计学院（ICD）及结构建造和设计学院（ITKE）通过一系列建筑原型，包括2012 ICD/ITKE展馆，

design and fabrication from the aerospace industry, the bottom up exploration in the architectural context has led to developments which become relevant to aerospace and might result in a reverse transfer of generated knowledge. A similar interdisciplinary process of investigation, abstraction, transfer and reverse transfer can also be observed in the case of biomimetic research. The investigation of biological precedents results in the abstraction of the underlying functional principles which can then be transferred into a technological application. During this process, methods or tools can be adopted for biological investigations and potentially generate valuable insights about the biological role model through a multi-disciplinary lens.

These methods allow for a reciprocal bottom up investigation of materialization strategies and biological lightweight construction principles. The explorations of new usage scenarios for materials and processes often expand the transferability of biological concepts as their complexity often exceeds the achievable level of differentiation with the given technology. In the case of the presented work it is the combination of robotic fabrication strategies and fiber composite materials and their transfer into the realm of architecture that allows a new degree of structural

Leichtbau BW复合表皮结构以及2013-14 ICD/ITKE展馆成功测试并应用了本文所介绍的设计方法。这些项目强调了研究纤维复合材料在建筑中先进的设计和建造策略有助于实现性能化的建筑设计品质。这些项目在建造过程中的资源节约很好地应对了建筑领域当前及未来需要面对的主要挑战之一。

通过进一步的研究，包括对前沿建造设计策略的探索以及对现有工业建筑应用发展流程的丰富和细化，这一系列的建筑原型还会继续。未来的研究方向包含将气候调节能力纳入几何生成的设计驱动，以及对材料组织策略的深化控制，这将提升功能整合以及系统性能。这包括通过立面方案和驱动策略的整合发展出综合的建筑体系。建造过程中的实时反馈发展出适应性建造策略，它模糊了设计与建造之间的区别。其余轨迹将通过大型复合材料的现场建造策略，努力克服因机器人工作空间导致的尺寸限制。所有这些理念都强调了规模化的性能整合，这将提升结构待开拓的潜力。

differentiation and thereby opens up potentials to implement biological principles in construction.

In the ICD/ITKE Research Pavilion 2013-14 the parallel investigation of coreless filament winding methods and biological fiber composite structures allowed the multi-disciplinary team of biologists, paleontologists, architects and engineers to develop a modular lightweight fiber composite building system. Its potential for architectural applications was tested with a full scale prototype.

During the investigation of functional principles in natural fiber composite structures the protective hind wing shell segments (elytra) in flying beetle species proved to be a versatile role model for lightweight construction in biological fiber composite structures. Flightless beetle species developed solid composite shells which are only perforated by thin channels for liquid (haemolymph) and gas transportation (trachae system). In contrast to these ground living species flying beetles need to maintain the structural capacity of their shells for protective reasons, but at the same time minimize the required energy to lift their body weight into the air. These seemingly contradicting fitness criteria were negotiated within the evolutionary development of flying beetles elytron morphology through the adaptation of the internal transportation channels into larger cavities. The connection between the outer and inner layer of the elytra was reduced to interstitial column like bracing structures (trabeculae). The variations in morphology and distribution of these trabeculae are linked to the local structural requirements within the global structural system. Through SEM scans and micro computed tomography methods a comparative study of multiple flying beetle specimens could be conducted. The result was a catalogue of abstracted morphologic and material organization principles. Based on these principles a modular fiber composite construction system was developed that simultaneously incorporates aspects of robotic fabrication, material behavior and structural considerations through a computational design tool.

Unlike the ICD/ITKE Research Pavilion 2012, in which the fabrication setup was tailored to create a single fiber composite shell, the fabrication of differentiated components requires and flexible robotic setup that could adapt to various geometries and unique winding syntaxes. A coreless filament winding system was developed that utilized a dual robotic setup [7]. In this system each robot was equipped with a 2m diameter rigid steel truss effector. These effectors provide a lightweight stable structure to attach various wrapping frames. The effectors and wrapping frames needed to be rigid enough during wrapping to resist the tensile forces applied by the tensioned fibers. The two robots had to be synchronized to maintain the distance and orientation between the wrapping frames as to not compromised the developing composite geometry either by releasing the tension on the system therefore allowing the fibers to deform or adding excessive

2013-14 ICD/ITKE研究展馆：机器人建造的碳纤维构成了建筑单体组件
ICD/ITKE Research Pavilion 2013-14 : Robotically fabricated fiber composite building component

tension to the fibers at the risk that the fibers might break. To maintain this synchronized movement one robot was designated as primary, and used to feed motion instructions for the winding process, and one robot was secondary, charged with following the relative change in location and orientation of the primary robot. The two robots, synchronized together, created a twelve axis kinematic system.

To further reduce the required formwork needed for the coreless winding system the winding frames were constructed from a small kit of reusable parts. Each component was geometrically determined by two non-planar polygons that each had between four and seven vertices. A clamping mechanism for the vertices was developed that had several degrees of freedom to adjust the X, Y, Z position and A, B, C orientation of each vertex plane. Taking advantage of robotic precision, an assembly process was developed to maneuver each robot into place so that each vertex clamp could be attached accurately in reference to the base effector. Upon each clamp was attached a five axis milled wooden guide that set the position and rotation of wrapping points on adjoining polygon edges. As much information was embedded into the geometry of these guides as possible to reduce the overall mechanical complexity of the adjustable wrapping frames. The integration of a robotic assembly process and the use of a reconfigurable kit of parts meant that accurate and robust wrapping frames could be created that required no additional surveying or digitization.

With an assembled set of frame geometries as well as synchronized robotic positions and movements, the wrapping of differentiated composite components is possible. Between the two robots is a stationary fiber source. A glass or carbon fiber roving component was impregnated using a dip type resin bath system [8]. The two synchronized robots rotate the winding frames into position so that the fiber source passes around each winding control point sequentially. Initially fibers are connected from one frame to the other in straight alternating lines. As more fiber are wrapping around the rotating frames fiber begin to cross and press one on the other. Eventually the interaction of the fibers forms a complex hyperbolic surface. Several layer of glass and carbon fiber layers are added generating the general form of the component and the specific structural reinforcing respectively. Each winding code, structural reinforcement layer, component geometry and connection details are coordinated with adjacent components in the global system.

The design and fabrication of each component required an integrated computational design method (ICDM). In the ICD/ITKE Research Pavilion 2013-14, this was implemented in several ways. First, it was implemented as a generative global design tool which integrated all the design constraints, objectives and variables. For this a physics based modeling system was implemented which abstracted design parameters as physical forces applied to a particle spring system. This system was used to control the individual component geometry within the larger global design of the pavilion. Any force acting on a component also acted upon it neighbors. Iterative geometric solutions could be represented by a state of global equilibrium between the forces. A second implementation of the ICDM is for simulation. Robotic fabrication processes were simulated to test the kinematic solution of the twelve axis robotic setup. Component geometries which were too large or exceeded the range of motion of the robot setup would be adjusted per the generative global design tool. A structural simulation was also integrated into the design loop allowing each iteration to be evaluated and adjusted for structural performance. Finally, an analytical process was utilized through the ICDM which allowed design iterations to be generated and evaluated through the filter of the performative design criteria.

As a result a full scale demonstrator was designed and constructed consisting of 36 geometrically unique components the largest at approximately 2.6 m in diameter and weighing 24.1 kg. Each component was wrapped without the need for prefabricated formwork or costly molds. Through the robotic fabrication process, differentiated fiber arrangements were created that responded

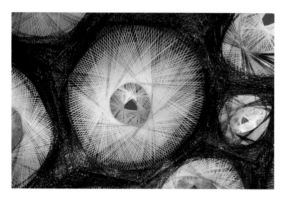

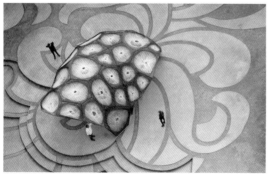

2013-14 ICD/ITKE研究展馆
ICD/ITKE Research Pavilion 2013-14

to various design and structural considerations. The overall demonstrator covered an area of 50 m² with a volume of roughly 122 m cubed. The pavilion was extremely lightweight weighing less than 600 kg in total.

The methods described in this article are successfully tested and developed at the Institute for Computational Design (ICD) in collaboration with the Institute of Building Structures and Structural Design (ITKE) through a series of architectural prototypes, including the ICD/ITKE Research Pavilion 2012, the Leichtbau BW Composite Surface Structure and the ICD/ITKE Research Pavilion 2013-14.These projects highlight the performative and architectural design qualities achievable through advanced design and fabrication strategies for fiber composites in architecture. The extraordinary reduction in resource utilization for their construction addresses one of the current and future central challenges for architecture.

This series of architectural prototypes will be continued through further research projects including explorations in cutting edge fabrication and design strategies as well as refinements and enrichment of existing processes for development towards industrial architectural applications. Future investigations will include the climate modulation capacity as a design driver for geometric articulation as well as further control of material organization strategies which increase the functional integration and performative capacity of the system. This includes the development of an integrated building system through the incorporation of façade solutions and actuation strategies. Developments in live feedback during the fabrication process would allow for adaptive fabrication strategies which start to blur the distinction between design and fabrication. Other trajectories strive to overcome the limitations in size incurred by the robotic workspace through on site fabrication strategies for large scale fiber composite parts. All of these ideas highlight the as yet unexplored future potentials for these structures to be enhanced by further performance integration and adaptation in scale.

鸣谢 / Acknowledgements:

本文中所提到的项目都是在计算设计到机器人建造的过程中通过多个团队合作来实现的，作为本文作者，我们想借此机会感谢所有参与过这些项目的研究人员，他们包括了：参与过2012ICD/ITKE研究展馆的斯图加特大学计算设计学院的史蒂芬·理查德，托比斯·斯科文和阿吉姆·孟格斯教授，斯图加特大学结构建造和设计学院的里卡多·拉玛格拉，弗里德里克·魏玛和简·尼普斯教授；参与过2013-14研究展馆的莫瑞斯 • 多斯特曼，马歇尔 • 普拉多和阿吉姆• 孟格斯教授，斯图加特大学结构设计学院的斯蒂妮·帕拉索和简·尼普斯教授；以及来自图宾根大学的生物学家们。

The authors would like to emphasize the collaborative character of the projects discussed here from a computational design and robotic fabrication project. Thus they would like to foreground the contribution of all involved researchers from the Institute for Computational Design (ICD) and Institute of Building Structures and Structural Design (ITKE) at the University of Stuttgart, who include for the ICD/ITKE Research Pavilion 2012 Steffen Reichert, Tobias Schwinn, Professor Achim Menges (ICD) Riccardo La Magna, Frédéric Waimer, Professor Jan Knippers (ITKE) and for the ICD/ITKE Research Pavilion 2013-14 Moritz Doerstelmann, Marshall Prado, Tobias Schwinn, Professor Achim Menges (ICD) and Stefana Parascho, Professor Jan Knippers (ITKE), as well as the collaborating biologists from the University of Tübingen.

参考文献 / References：

[1] Palucka, Tim and Bensaude-Vincent, Bernadette. "Composites Overview." Pasadena, CA: California Institute of Technology, 2002. Web. 02 Mar. 2015

[2] Grave, Guido, et al. "Simulation of 3D overbraiding—solutions and challenges."Second World Conference on 3D Fabrics and their Applications, Greenville. 2009.

[3][4] Uhlig, K., et al. "Development of a highly stressed bladed rotor made of a cfrp using the tailored fiber placement technology." Mechanics of Composite Materials 49.2 (2013): 201-210.

[5] Reichert, Steffen, et al. "Fibrous structures: an integrative approach to design computation, simulation and fabrication for lightweight, glass and carbon fibre composite structures in architecture based on biomimetic design principles."Computer-Aided Design 52 (2014): 27-39.

[6] Dörstelmann, M., Parascho, S., Prado, M., Menges, A., Knippers, J: 2014, Integrative Computational Design Methodologies for Modular Architectural Fiber Composite Morphologies. In Design Agency [Proceedings of the 34th Annual Conference of the Association for Computer Aided Design in Architecture (ACADIA)], Los Angeles, pp. 219–228.

[7] Prado, Marshall, et al. "Core-Less Filament Winding." Robotic Fabrication in Architecture, Art and Design 2014. Springer International Publishing (2014): 275-289.

[8] Miaris, Angelos, and Ralf Schleojewski. "Continuous impregnation of carbon-fibre rovings." JEC composites 56 (2010): 75-76.

松江艺术产业园区的红色清水砖墙,2015
The Red Plain Brick Wall of Songjiang Art Campus, 2015

New Craftsmanship Using Traditional Materials
传统材料的数字新工艺

Philip F. Yuan, Hyde Meng, Zhang Liming / Digital Design Research Center, CAUP, Tongji University
袁烽，孟浩，张立名 / 同济大学建筑与城规学院数字设计研究中心

数字设计革新

从20世纪90年代开始，工业化生产出现了从机械化技术向数字化技术的过渡。这样的转变对建筑设计和建造产生了深远的影响，并不断启发建筑师对本体的思考和创新。而如今，使用数字化技术实现定制化生产也成为了信息时代的驱动力量。工业4.0时代的关键是原材料的信息植入，数字工厂中每一份材料都有准确的参数信息定义自己在产品中的角色 [1]。在大规模生产中经济地制造出非单一的建筑产品将是数字设计的革新范式。

数字设计的范式转型还体现在建筑师对建筑参数本质的再审视。帕特里克·舒马赫通过参数符号学强调参数化建筑的基础是建筑自组织系统理论，建筑师的核心竞争力是对复杂的建筑信息系统的设计。通过运用系统化的设计方法和缜密的数字逻辑定义，排列、组织与设计建筑参数，是实现建筑与内外部环境协调共生的核心；也是将建造逻辑脱离纯粹的几何形式，与材料属性和建造行为密切联系的关键 [2]。

传统材料的数字建构

纵观建筑风格与思潮的发展历史，不难发现的一点是：风格与思潮的转变只是表象，其背后设计方法与设计工具的变革才是真正影响

The Digital Design Revolution

Since the 1990s, within industrial production there has been a transition from analog technologies to digital technologies. This transition has exerted a profound influence on architectural design and building fabrication, and has inspired architects to rethink their design methodologies. Today, customized production achieved by using digital technologies has become one of the driving forces in our Information Age. The key is to embed information into raw materials. Each material used in digital production can be defined very precisely in terms of its role in the final product [1]. The economic mass production of heterogeneous building products is one of the characteristics of the revolutionary new paradigm in architectural design [2].

This new digital design paradigm is also manifested in architects' re-examination of the nature of building parameters. Patrik Schumacher argues that one of the fundamentals of digital architecture lie in autopoetic design methodologies achieved through parametric techniques, and that core competitiveness of architects lies in the design of complex building information systems. To arrange, organize and design building parameters using systematic design methods and meticulous digital

建筑范式与建构文化的决定性因素。譬如文艺复兴背后的透视学，巴洛克建筑背后的切石法，还有现代主义背后克林·罗的"透明性"概念驱动下的轴测画法。传统材料的数字化建构就是试图通过计算机技术与数字建造工具，使得材料的建筑表现力向性能化美学进化，通过对材料物理性能参数的分析，设计与优化，提高建筑与场地、环境气候甚至使用者行为方式的积极互动[3]。

传统材料的固有特性决定了建筑单元体加工建造的方式，单元的几何形态，单元之间的连接方式、间距、角度改变着整体材料的性能表现[4]。比如竹木材料主要考量抗拉与抗压性能、弯曲变形能力、结构跨度与截面尺寸、环境作用与形变的关系等；砖石则侧重于硬度、规格、导热性、吸水性等；陶土的重点属性则包括吸水性、透气性以及塑形能力、成型速度等。同时，竹木的取材、定型以及连接件的设计和榫卯的构造，砖体的烧制过程，砌筑方式以及砂浆的选取和使用，还有陶土的翻模工艺和烧制方法等工艺都是数字化建造的理论基础。针对材料特性所产生的建造信息串联设计和施工，同时为设计本身赋予了更深层的含义。建筑师在设计中必须考虑材料和结构在物理上的可操作性，同时将建造的时间和空间次序包含在内。这一套建造的"预审"机制是数字化建构的核心价值[5]。

基于材料性能的设计建造方法

基于材料的设计方法关注建筑自身性能与材料物质属性之间的紧密联系。通过以材料研究为对象，以探索新工艺为目标，以数字化设计为方法，以数字化建造为手段，创造从生形，优化到建造的一体化建筑设计工艺流程。

该设计方法从材料的性能性研究和数字化模拟入手，通过在建筑软件中赋予设计材料属性，模拟材料在重力、外力以及自然环境下的物理变化，在概念设计的过程中直接引入对建筑性能水平和建构方法的思考。以材料的物理参数为基础的建筑生形设计是一种自下而上的原型设计，具备了多目标参数性能优化的条件[6]。优化设计软件整合包括原型的结构性能与环境性能参数的多目标设定，对不同的材料参数进行量化处理，权衡比重，制定综合评价机制，设定优化的适应性函数，通过遗传算法按照优胜劣汰的自然原理进行迭演化，最后以微差表达的形式达到对材料单体和建筑原型进行设计优化的目的。

对建筑材料的设计考量也使数字建造过程更为直接。建筑师可以在设计中模拟建造过程，针对材料特性选择最合适的建造逻辑，材料经过性能优化后产生的形式微差与变化往往需要建筑师量身设计最有效的数字加工工具。以砖的数字化建构为例，机器人堆砌技术相比传统堆砌技术的优势在于基于数据信息的精确堆砌。传统堆砌方法与砌体的形态有直接关系，如方形的砖块砌筑方式即为横平竖直的砌筑，而机器人堆砌完全是基于数据信息，打破了砌块本身形状的限制，可以实现极为多样而精准的砌筑。因此砌筑工具端的开发需要着重于工具头的空间定位技术和砂浆涂抹技术[7]。而竹木原型建造的数字工艺则主

logic is the key to achieving a symbiotic relationship between architecture and the exterior environment. According to Schumacher at least parametric semiology not only liberates construction from simply following geometrical definitions, but closely connects it with material attributes and fabrication procedures.

Digital Tectonics Using Traditional Materials

An overview of developments in architectural design and thinking will reveal that although changes in forms and styles are superficial, the revolution in design methods and tools is the decisive factor in actually influencing architectural paradigms and tectonic culture, such as the development of perspective during the Renaissance, stereotomy during the Baroque period and isonometric drawing techniques inspired by Colin Rowe's concept of "transparency" within modernism. Digital fabrication using traditional materials attempts to introduce a new performance-based logic to the architectural expressiveness of building materials by using computer technologies and digital fabrication tools, in order to improve the relationship between buildings and their physical context, environment/ climate and users' behaviors through the analysis, design and optimization of a material's physical parameters [3].

The inherent properties of traditional materials determine the manufacture and fabrication methods of building elements. The geometrical form and the angles, space and connections between elements change the overall material performance [4]. For instance, timber is investigated by its tensile strength, compressive properties, bending deflection, structural span and sectional size; bricks and stones are measured by their dimension, rigidity, thermal conductivity and hydroscopicity; while the main physical characteristics of ceramics include permeability, porosity and plasticity. Meanwhile, traditional techniques such as the selection and sizing of timber members, the design of their connections through mortise and tenon joints, and likewise with bricks, the firing process, layng techniques, and the composition and application of mortar, become the theoretical basis for digital fabrication. Construction information related to material properties link design to fabrication, and have fundamental implications for architectural design itself. During the design process, architects should consider the physical feasibility of both materials and structure within the construction timeline. This set of preliminary concerns lies at the core of digital fabrication [5].

Digital Design and Fabrication Based on Material Performance

Material performance-based design approach emphasizes close relationship between building performance and material property.

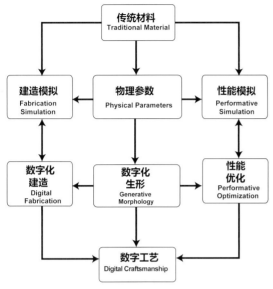

技术路线图
Design Strategies

要针对数字设计中复杂多样的构造节点进行相应的切割工具开发。数控切割相比传统切割工具的优势在于可以实现复杂曲面的切割,尤其是需要精确定位的空间曲面。这类工具端的开发将着重于切割工具的材料、形式、运动速度与被切割竹木纤维排布之间的关系,以达到良好的建构效果。

传统材料的数字实践

传统材料的建造实践价值不仅限于文化价值,当建筑师寄予传统材料的更多性能特征以及特定的建造工法,传统材料的意义将被重新被定义。更重要的是传统材料所赋予的鲜明地域特征,毋容置疑地给建筑设计注入了丰富的地域元素特色。近年来,笔者致力于对传统材料的数字建造再创作,从四川成都非物质文化遗产园兰溪庭的水纹外壁,到松江艺术产业园区的红色清水砖墙,到卜石艺术馆的公共空间,传统材料正在经历一场新的建造工法以及形式意义的试验。

在松江艺术产业园区的项目中,建筑师通过参数化设计,将传统的"丁-顺"红砖砌法通过非线性的逻辑加以重构,简单的定位与控制方式让工人通过简单的学习就可以加以营造。建筑墙体呈现出如织物般柔软,渐变的效果。通过数字技术的优化,砌体最后采取了八种进退关系进行施工,既满足了立面丰富的纹理效果,也保证了施工的合理性。在施工现场,建筑师为工匠提供精确的定位模板作用于砖块的砌筑,虽然精准性是手工级别,但仍然在快速低成本的施工过程中最大化的保证了设计初衷。

卜石艺术馆多层次的艺术空间通过面向庭院的中央交通空间而得以组织,老建筑原有的单调线性平层模式被一个重新置入的非线性混凝土空间所重塑。艺术馆中简单的竖向和横向交通动

Taking material research as object, craftsmanship exploration as objective, digital design as methodology, and digital fabrication as tool, an integrated architectural design procedure from forming to simulation and construction is emerged.

This design approach starts with research into material performance and with digital modelling. By modelling the physical changes in materials under the influence of gravity, external forces and the natural environment, and by incorporating material properties into design via architectural design software, architects will take the performative aspects of building materials and their construction logic into consideration during the conceptual design phase. An understanding of architectural morphology based on the physical parameters of materials offers us a bottom up approach to design, which is capable of accommodating multi-objective evolutionary optimization. The integration of optimization design software involves the multi-objective testing of the architectural prototype according to structural and environmental parameters. It calculates and quantifies the parameters of different materials, establishes a comprehensive selecting mechanism, sets up synthesized fitness functions, and performs an iterative evolutionary process using genetic algorithms based on natural selection – all in order to achieve the overall optimization of architectural prototypes through the micro differentiation of each material component.

Design according to material performance also makes digital fabrication more direct. The architect is able to model the fabrication process during the design phase and to select the most appropriate fabrication strategy based on material properties. As a result the architect is able to select the most effective digital fabrication tool for the construction of the varying material components. Taking the tectonic properties of brick as an example, the advantage of robotic masonry construction over traditional masonry construction lies in the accurate manipulation of digital information. Traditional masonry construction relies directly on the form of the brick. For example, a square brick can be piled up horizontally and vertically. Robotic masonry construction, however, is not limited by the form of brickwork, and is able to realize extremely diverse and complex patterns of masonry. Therefore, the development of robotic masonry construction requires accurate spatial positioning and intelligent mortar composition [7]. Meanwhile in robotic timber fabrication, an appropriate cutting and milling tool needs to be developed for complex joints and connections. CNC robotic fabrication is superior to traditional methods due to its capacity to mill complex surfaces, and spatial geometries that require accurate control of the tools. The development of such kinds of digital tools should pay close attention to factors such as materiality, geometry and speed based on the distribution and properties of fibers within the

线在这个空间中交汇且彼此影响，相互之间边界的模糊带来了空间的折叠与揉合[8]。带来连续一体的几何相位切换的同时是对人运动体验的重新定义。非线性的度量和几何分解成为设计实现的核心所在，计算机中抽象的多向度曲面被分解为可控制的CNC可加工片段，而立体的拼接则通过精确的定位加以控制。曲面化的造型在符合其几何规律的前提下被转化为线性的加工逻辑。传统二维图纸无法实现的空间表达转化为建造逻辑的表达，数字化的建造过程使得设计思路从始至终加以贯彻。

传统材料的性能化建构

依托同济大学数字研究中心强大的数字建造实验平台，笔者试图将实践中遇到的问题和迸发的想法在研究中寻找出解决办法和答案。实验室通过对材料科学、计算机科学以及机器人建造技术的跨界结合，探索建筑材料的性能化特质，建立一种从数字化设计到建造的一体化设计方法。通过对数字化模拟技术与机器人建造技术的整合对接，将建造工艺、材料性能以及环境模拟融入到建筑设计和建造的全程中。研究从砖石、竹木、陶土等传统材料的物理属性和建造方法研究出发，建立针对建筑生形的材料工具包，结合数字建造和机器人技术，通过系统的数字设计建造方法，对传统工艺进行性能化革新。

卜石艺术馆，2013
Jade Museum, 2013

chosen material, so as to achieve a satisfactory tectonic result.

Digital Fabrication Using Traditional Materials

Traditional craftsmanship should not see only in terms of its cultural value. The significance of traditional materials will be redefined as the architect pays closer attention to its performative characteristics and fabrication logics. What's more, the unique geographical features embodied in traditional materials bring rich regional characteristics into architectural design. In recent years, the author has devoted himself in his architectural practice to robotic fabrication using local materials. From the exterior waving surface of Lanxi Curtilage in the intangible cultural heritage garden in Chengdu, Sichuan, to the plain red brick wall of the Songjiang Art Campus, and the public spaces of the Jade Museum, Shanghai, traditional materials have been subjected to morphological experiments to achieve a new creative form of architectural tectonics.

In the Songjiang Art Campus, the architect has reconfigured the traditional "Ding-Shun" brick construction method (Flemish bond, cross bond) with the nonlinear logic of parametric design. Simple positioning and measuring methods enable the workers to understand the construction process. The brick wall creates an effect of a gradually folding vitiated texture like fabric. Eight convex-concave relationships have been adopted in the construction of the masonry, which not only create different textured effects for the building elevations, but also ensure the rationality of construction. On the construction site architects provide customized positioning tools to the builders to determine the location of each brick. Although the accuracy depends upon the level of the craftsmanship, nonetheless it still guarantees the initial design intention during the course of construction while maintaining the speed of construction and keeping within the budget.

The multi-storey art space of the Jade Museum is organized around a central circulation stairway opening directly off the courtyard. The original monotonously linear floor to floor model of the old building is reconfigured through the insertion of a nonlinear concrete space, where vertical and horizontal movements intersect and interact. The ambiguity of the interfaces between the different spaces creates a variegated folding and blending effect [8]. The movement of the visitors is re-defined while there is a continuous variation of integrated geometric profiles. The integration of nonlinear metrics and geometry becomes the key to the design. The multi-dimensional curved surfaces are abstracted and decomposed into controllable CNC milled fragments while the three dimensional field splices are controlled by accurate positioning techniques. The curved profile is translated into linear construction logic

机器人绸墙

本项目基于单元砖体均匀分布的材料特性，通过改变砖体之间的连接方式及空间发展方向，突破了长久以来其由本身的形态刚性及结构受压特性所产生的空间结构极限。设计采用砖体的形式语法进行生形并对墙体的结构性能进行优化，通过遗传算法对产生结果进行迭代筛选，使得砖这种古老材料在本质意象和传统做法中寻找结构性能上的更多可能性。机械臂的空间精准定位使得这种预设走向实际建造，保证了由于空间和条件的局限而选定的分段搭建，在多段墙体连接后能够形成一段稳定连续而具有强烈视觉冲击力的砖墙序列。与此同时机械臂工作站的综合性流水线工作模式，大大提高了现场施工的效率和准确度，保证了数字砖墙结构性能的完整度。

边锋椅

边锋椅通过建筑几何生形逻辑模拟人体肢体构造，从行为方式和材料性能方面探索人体与家具以及建筑之间的互动关系。设计通过对自由曲面起伏变化的理性操作，将座椅在人体工程学中的核心参数作为形态的迭代优化标准，通过人体坐姿的准确定位和记录，个性定制座椅支撑，坐垫扶手以及靠背的形态关系，通过曲面的光滑过度来达到最舒适的使用体验。在建造过程中，通过对几何形体的准确分析，并结合对于材料自身尺寸，纹理以及强度的综合考虑，通过切片细分和机器人五轴铣洗技术最大化的展现了木材的柔性和韧性，在实现人体触感与木材肌理完美贴合的同时消除了传统制作工艺中个体偏好与批量生产之间的矛盾。

反转檐椽

该项目抽取了传统木构建筑中"檐椽"这一元素，希望通过新的结构性能模拟优化和机器人加工手段对这一传统元素的结构性能进行优化与重新演绎。《营造法式》中对檐椽的出挑比例有明确的规定，设计师通过基于参数化设计平台的结构计算软件Millipede和遗传算法优化运算器Galapagos对这一出挑比例进行了验算与优化，结合由三根杆件形成的基本的三角形自支撑单元，设计了一套新的结构体系。整个结构装置由七个单元组成，虽然这七个单元遵循同一种逻辑，但杆件的长度，倾斜角度和搭接位置各不相同，而机器人的精确加工能力能够很好的解决这些问题，使得本装置的结构意图得到完美呈现。

机器人制陶

陶土打印是数字设计实验室对传统材料建造方式的又一挑战，项目提出这样的命题：机器人是否能根据材料的内有属性，做出针对性的区别处理。设计师通过数字模拟技术，在机械臂运动的逻辑、技能和机制上将制陶工艺进行扩展，以框架和规则定义建造的特性，并直接与陶土的性能联系在一起。在打印过程中，陶土的透气性、水合程度、塑性能力与打印工具的挤出

under the premise of satisfying the geometric principles of the design. The spatial expression that would have been unrealizable through traditional two-dimensional drawings is transferred to the sequential logic of the fabrication process, so that digital modelling ensures the implementation of design initiatives throughout the whole construction process.

Performative Tectonics with Traditional Materials

Through the powerful digital fabrication facilities of the Digital Design Research Center at Tongji University, the author has tried to look for solutions to the questions encountered and challenges faced in architectural practice. The laboratory explores the performative features of building materials and tries to establish an integrated approach from digital design to fabrication through the interdisciplinary integration of material sciences, computer science and robotic technology. Digital craftsmanship, material performance and environment analysis are incorporated into architectural design and fabrication through the integration of digital modelling and robotic fabrication technologies. The research focuses on the physical properties and fabrication methods for brick, wood, pottery clay and other traditional materials, establishing design tools and material libraries for architectural form generation, while replacing traditional

基于传统材料的机器人性能化建构
Performative Robotic Fabrication Based on Traditional Materials

速度、机械臂自身运动的速度、加速度甚至加工空间的温度和湿度都息息相关。只有通过对各参数严格的控制和模拟，并根据设计的形态和建造的过程不断的协调各参数间的作用，才能得到优化的结果。建筑师通过在数字设计过程中将材料的性能以参数形式传递到建造工具，融入到构造逻辑中，使工具和材料共同反应，拓展了数字化建造的深度和广度。

数字新工艺

如今，参数化形式范式转型正在深度转向了对于参数意义的全新诠释。数字逻辑下，我们可以更加精确地运用系统化的设计方法，探索材料的性能表现、建构文化以及人的行为方式等具体化的参数意义。"在地性"不仅包含了建构文化方面的意义，还包含了建筑与气候，建筑性能与场地环境，以及地方化的行为习惯等诸多方面内容。数字时代下，再谈地域主义具有更加丰富的伦理意义。新时代的建筑师需要将建筑实践与中国传统文化语境结合，通过数字设计和建造方法，开发本土建筑材料的性能美学，结合现代科技与地域文化，一起推动中国数字新工艺的革新。

机器人制陶
Robotic Ceramic Printing

craftsmanship with performative digital fabrication.

Robotic Wall

This project breaks through the spatial and structural constraints caused by the rigidity of the form of the brick and its loading capabilities - which has constrained brick construction for a long time - by changing the bonding logic between the bricks and their spatial orientation based on the physical features of regular bricks. The design optimizes the structural performance of the brick wall using aggregative digital techniques. The iterative evolution of the design is carried out by using genetic algorithms to search for possibilities of maximizing the structural performance of the material beyond traditional approaches. Accurate positioning of the robotic arm ensures the deployment of this design approach through to actual fabrication. Meanwhile, an integrated assembly workstation with a robotic arm and automatic brick supply system greatly improves the efficiency and accuracy of site fabrication and ensures the overall structural performance of this digitally fabricated brick wall.

Edge Chair

Edge chair simulates human body structure through architectural morphologies, and explores interactive relationship among human body, furniture and architectural space from the aspects of material performance and behavior preference. With proper control on the change of undulating surface, the design uses core parameters of chair in human engineering as optimization criteria for the iteration of geometrical form, most comfortable sitting experience can be achieved through smooth transition of surface's curvature, proper positioning and recording of human gestures, and personalized chair brace, handrail and backrest with specific geometry tailoring. Flexibility and tenacity of timer is embodied at maximum during fabrication process after accurate analysis of geometric form, comprehensive consideration of material dimension, texture, and rigidity. Through the employment of sectioning technology and robotic milling, It not only realizes the perfect combination of human sensation and timber texture, but also removes conflicts between individual preference and mass production in traditional craftsmanship.

Reverse Rafter

The traditional wood rafter is used in this project, and then processed and reconfigured using new structural performance modeling tools as well as robotic fabrication. In Yingzao Fashi there are clear instructions for the proportions of projecting rafters. The designers have checked and refined the proportions of these rafters parametrically by using the structural performance software, Millipede, and the Genetic Algorithm based optimization

机器人制陶
Robotic Ceramic Printing

tool, Galapagos, establishing a triangular self-supporting unit through combination of three timber members, so as to produce a new structural system. The whole structure is composed of seven units which follow the same logic. But the problem is that each individual member must be different in terms of length, angle of inclination and joint. Precise robotic fabrication can solve this problem and express the design intent of this structure perfectly.

Robotic Ceramic Printing

Ceramic printing is another challenge to traditional material craftsmanship. This project asks whether a robot can provide an adaptable fabrication strategy based on the inner properties of materials. Through digital modelling, the designers have expanded the logic of ceramic fabrication through the interpretation of a robot's manufacturing logic, skills and operating mechanism, so as to define fabrication logics with preset constraints and regulations, and establish direct connections with the properties of clay. During the fabrication process, factors such as air permeability, degree of hydration and the plasticity of ceramic materials are carefully calibrated so as to relate closely to the extraction velocity of the printing tool, the speed and acceleration of the robot arm as well as temperature and humidity within the physical space. Optimized results can only be achieved through strict control and modelling of various parameters, and by consistent coordination of the feedback between the design outcome and fabrication process. The designers transfer the material properties into the design of the fabrication tool and the logic of the construction process so that the range of the fabrication process can expand through the combined virtual and physical manipulation of material and tools.

New Digital Craftsmanship

Today, the paradigm of parametricism has shifted its meaning in terms of the use of parameters. Using the logic of numbers, architects can apply integrated design methodology to the discovery of the specific definition of various parameters such as material performance, architectural tectonics and human behavior. Regionalism, meanwhile addresses not only local craftsmanship, but also local climate, site information and local culture and behavior. The discourse of regionalism in the digital age has a broad ethical significance. Architects in this new age combine advanced architectural practice with traditional Chinese culture, developing performative local building materials and adaptive forms of architecture through digital design and fabrication, exploring the new possibilities of the integration of parametricism and regionalism through new digital craftsmanship.

参考文献 / References：

[1] Mario Carpo, 'Digital Darwinism:Mass Collaboration, Form-Finding, and the Dissolution of Authorship', Log 26, Anyone Corporation, New York (2012): 97–105.

[2] Patrik Schumacher:The Autopoiesis of Architecture: A New Framework for Architecture, Wiley (2011): 5-8.

[3] Michael Hensel, Achim Menges. Versatility and Vicissitude: Performance in Morpho-Ecological Design. Wiley Publisher. 2008(4): 120-125.

[4] 尼尔·里奇，袁烽，建筑数字化建造 [M]．上海：同济大学出版社，2012:107-108.

[5] Fabio Gramazio, Matthias Kohler and Jan Willmann, The Robotic Touch: How Robots Change Architecture, Park Books, Zurich (2014): 4.

[6] 徐卫国．数字建构 [J]．建筑学报，2009(1): 61-68.

[7] Bob Sheil, Ruairi Glynn, Fabricate: Making Digital Architecture, Riverside Architectural Press (2012): 120-121.

[8] 袁烽，葛俩峰，韩立．从数字建造走向新材料时代 [J]．城市建筑，2011(5): 10-14.

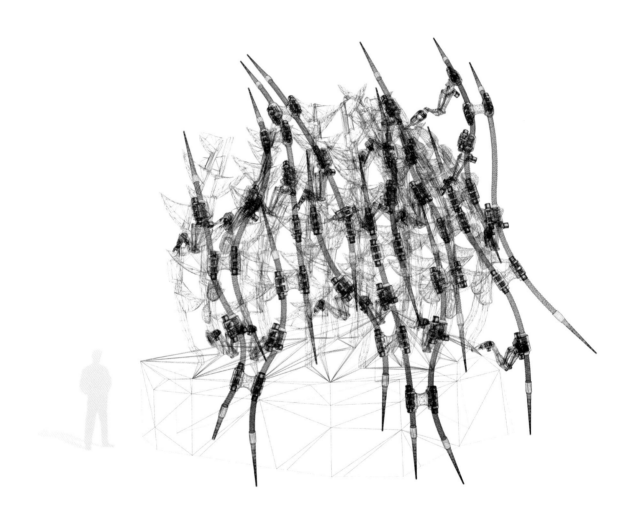

Machines for Rent
出租机器

Francois Roche, Camille Lacadee / New-Territories
弗朗斯瓦·罗氏,卡米尔·拉卡迪 / 新领域事务所

在小说、艺术和电影对于科幻题材的叙述中,机器人已经吸引了大众的想象力,本文是在幻想领域对于这种现象的探索,这种探索成为了推动实验的沃土。
Robots have captured the popular imagination like no other technology, fueling science-fiction narratives in novels, art and film. This text is explored as a phenomenon that inhabits its own visionary realm, becoming a fertile field for experimentation.

有很多机器,它们似乎能满足人们的需求,但事实上这些机器能做的事情远远少于其声称所能做的。在追求超然科学——解决想象领域事宜的哲学分支——的道路上,它们从来都不会暴露出其深层特性:无论是它们的血统或是它们幻觉般的外观,它们真实的品质或它们虚假的特点。这些机器是推测性的、虚构的,同时也具有准确的、高效的生产力,它们航行在从不知道是今天还是何时的世界,快乐地、天真无邪地,轻快地穿过21世纪无数的垃圾山。这些超自然科学的机器通过奇怪的设备进行系统化的连接,在不同的时间流向、不同的知识层面,越过我们所认为的荒谬领域的无尽的极限;在那里,不合逻辑的行为用极富逻辑的新兴设计和几何结构得到确认;在那里,输入和输出通过数学规则进行描述……

既不是对C.S.刘易斯式浪漫科幻的讽刺,也不是技术悲观主义或技术讥讽,这些机器停留在反乌托邦的限制上,又或者建立

There are many machines – so many desirable machines – that in fact pretend to do more than they are actually doing. In the pursuit of pataphysics – the branch of philosophy that deals with the imaginary realm – they never reveal their deep natures, whether it is their lineage or their illusionary appearance, their genuine qualities or their sham features. Simultaneously speculative, fictional and accurately and efficiently productive, these machines navigate the world of 'Yestertomorrowday', with happiness and innocence, passing briskly through the mountains of rubbish of the 21st century and beyond. These pataphysical machines articulate symmetrically – through weird apparatuses – different arrows of time, different layers of knowledge, but – more efficiently – they negotiate the endless limits of what we could consider the territory of the absurd, where illogical behavior is protocolized with an extreme logic of emerging design and geometry, where input and output are described by mathematical rules …

惯例性、必然性和稳妥性领域之间的限制。在这些领域，人们一般认为任何事情都是由规则或对规则的直觉和其他领域确定的。所有其他类似妄想、幻想的想象都是由其他领域的探索者们反馈回来的。

从基本观念上来说，机器通常通过替代手工、提高效率、速度以及转换的力量来体现技术作为手工的延伸。但是，若仅仅将机器限制在这一显而易见的客观维度，或者单纯的功能和机械范围内，限制在笛卡尔式的有关生产力的概念以及有关外观和事实的可见范围内，则显得过于天真。同时，机器可以用于生产人工制品和聚合物来满足多种功能需求，渗透进了与我们的身心相互依赖的环境和栖息地。从根本上说，在自然的任何地方，在所有交换过程的起源，在任何物质的相互作用过程中，机器都是活力的保证。机器和自然的共存使得它们实际上成为了一种身体的典范。这适用于所有流程、协议和装置，在这些流程、协议或装置中，瞬变的、互相影响的物质构成并同时影响所有物种，机器的身份和机器的输出既是客观的、又是主观的。

在寻求复合方法过程中，我们不能忽略"单个机器"的概念，"单个机器"的概念作为一种初步尝试，将机械装置以一种炼金术的模式融入叙事性的交易和演变之中。这种方法与轻率的批判或抨击强调用操作机器的不熟练工人代替技工的资本主义（这种自然结果便是没有工人的机械系统）不同。本雅明将这种转变描述为从定制化的生产转变为大批量生产。这与怀旧的浪漫主义完全不同，浪漫主义认为由于我们迷恋机器精致的"类似人造"的结构，所以单个机器才得以发展。这种"冲动的欲望"和极度痛苦的压抑意味着它们长期处于一种精神分裂的状态，在并置的生产与破坏之间徘徊不定。正面和负面的流程都是同一工业系统的产物。它们的起源是同质的，但它们所产生的附加效果却完全不同，同时，它们又都依赖于这种分裂的可能。

以下是一些例子，代表了这种叙事性产品的病态性策略。

版权 / Credits :

文字 Text © 2014 John Wiley & Sons Ltd.
图片 Images © New-Territories / Francois Roche / Stephan Henrich

Neither a satire of 'this and other worlds', nor a techno-pessimism or a techno-derision, these machines reside at the very limits of the dystopian or they constitute the limit between the territory of conventions, of certainties and stabilities, where it is comfortable to consider everything legitimated by an order, or an intuition of an order, and by other territories – all the other paranoid, phantasm-like imaginings reported back by travelers.

In a casual and basic sense, machines have always been used to elaborate technological operations as the extension of the hand, through its replacement, and improvement, its acceleration of the speed and powers of transformation and production. However, it seems very naive to reduce the machine to this obvious objective dimension, in a purely functional and mechanical approach – limiting it exclusively to a Cartesian notion of productive power, located in the visible spectrum of appearance and fact. In parallel, machines can produce artifacts, assemblages, multiple associations and desires, and can infiltrate the very raison d'être of our own bodies and minds that are codependent on our own biotopes or habitats. Fundamentally, everywhere in nature, at the origin of all exchange processes, in the transaction of any substances, they are the guarantee of its vitalism. Machines' coexistence with nature renders them in effect a paradigm of the body. This is true for all processes, protocols and apparatuses, where transitory and transactional substances constitute and affect simultaneously all species – where machines' identities and outputs are both object and subject.

In pursuit of this polyphonic approach, we cannot pass over the notion of 'the bachelor machine'4 as a tentative attempt to integrate mechanical apparatuses in a narrative of transaction and transmutation (in the mode of the alchemist). This is the opposite approach to a headlong critique or denouncement of capitalism that highlights the substitution of craftsmen with unskilled workers manning machines (the natural consequence of this now being a mechanical system without workers). Walter Benjamin described this shift as a move from the singularity of production to mass production. This contrasts with the nostalgic romanticism that bachelor machines evoke through our fascination with their sophisticated 'human-like' construction – their eroticism or even barbaric eroticism. The 'impulsive urge' and gut-wrenching repulsion they generate means they exist in a permanent state of schizophrenia, vacillating between the simultaneous potential of production and destruction. Both positive and negative processes are the product of the same industrial system; their genesis is symbiotic, and their collateral effect in diametric opposition. They are both dependent on this schizoid potential.

Here are a few examples of those pathological strategies of narration-production.

达尔文星门

使用说明：

站起来，直面你私家花园深处的幽灵！

租用这辆车，你可以从坐着的、平静的、昏昏欲睡的姿势，转变为站立的、头脑清楚的、清醒的姿势，以便可以勇敢地面对当下。

达尔文星门由光伏电池供电，其手臂可以在移动的过程中展开，从完全展开的相貌转变为你通常情况下会拒绝看的令人不悦的异位空间。

星门机器引入了不同起源和周期的两种结构之间的时间推移。作为一种探究时间流向的方法，它能够通过"把我传上飞船"这样具有韵律感的捷径，平息现代到后现代，后现代到数字时代，数字时代到机器人计算延伸之间的转换所带来的不理解引起的焦虑。这种旅行能够产生进化和/或回归轨迹。但这个机器最重要的是它可以作为一个矢量，使人最为有效地从任何事物均被拉平、分类和验证的区域中逃离出来，达到不确定的、非决定性的目的地点。

它首次用于"thebroomwitch"实验。

注意事项：

过于频繁使用该机器不仅可引起时间剥夺感，有时还可能产生永久性感觉。同时其还可能最终成为你在特定时间拒绝履行义务的一个很好的借口。你也可能感觉不到时间的流逝，这可能显著影响电机动作的同步性。滥用该设备可能严重危害心理健康，并极大影响用户的临时感知，特别是对当前状态的感知。最终，它可能造成记忆丧失。

相反，停留于当前时间过长（停留在某个时间或另一个时间）可能造成用户沮丧、产生愤世嫉俗的行为或其他病态痛苦。该机器对因此而造成的此类感觉、行为等不承担任何责任……

该机器不适用于法国建筑，因为法国建造使得其起源变得模糊不清。

Darwinian Star-gate

Instructions:

Stand up and face the ghosts in the depths of your private garden!

Rent this vehicle to transport yourself from a seated, peaceful, sleepy archaic body posture to a standing, lucid awakened position that induces bravery in those faced with the present.

Powered by photovoltaic cells, the Darwinian Star-gate's arms unfold on their way from a panoptical to a troubling heterotopic space that you would normally refuse to see.

The star-gate machine introduces the passage of time between two constructions of different origins and periods. As a strategy for questioning the orientation of the arrow of time, it is able to relieve the anxiety of misunderstanding provoked by the shifts from the Modern to Postmodern, Postmodern to Digital, and Digital to Robotic Computational extension, in a 'beam me up, Scotty' rhyzomatic shortcut. The travel could take an evolutionary and/or regressive trajectory. But – overall – this machine is most efficiently used as a vector of discovery that reaches a point of uncertainties, of non-determinism, to escape from a zone where everything has already been flattened, classified and validated.

Its first use and development was for 'thebroomwitch' experiment.

Precautions for use:

Using the vehicle too often might cause a sensation of time deprivation and sometimes immortality, but also ultimately a good excuse for denial of your responsibilities in any given time. You might also lose the sense of time passing, which can significantly impair your synchronized motor actions. Abuse of the device can be extremely dangerous for mental health and may seriously affect the user's perception of time, especially in regards to the notion of a specious present. Ultimately, it can cause memory loss.

On the contrary, overexposure to the present time (staying in one time or another) might cause the user depression, cynical behavior or other pathological distresses, which the vehicle shall not be held responsible for...

The device does not work for French architecture, which already confuses its origins.

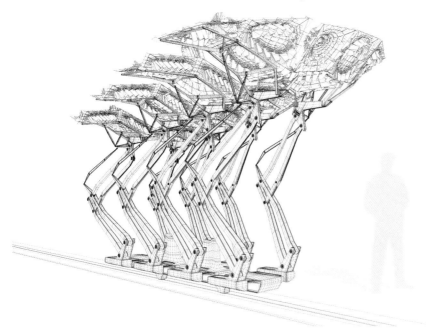

杀伤性漫游者

使用说明：

租借该机器可使得您勇敢地面对危险，并从任意"无人地带"上运回腐烂的物种及生物质。

可对该机器进行改造用来收集其他材料。如你可以按照该机器的初始状态归还，那么你可以进行所有机器人终端、铰接臂、腿及尖端的机械"调试"。

该机器收集可回收的任何材料以使用于新的生产过程中。这赋予被污染区域中的废弃物及垃圾第二次生命。被污染的区域，如基础设施严重坡坏的战后区。

与此同时，传说及童话从那些废弃的场景被提取出来，犹如在"潜行者"（切尔诺贝利核电废墟的探索）实验中触碰未知物一般。需要注意的是这些创造物的反作用。

注意事项：

该机器原配备有极高的自我评估感应设备及有致盲危险的部件。这些是其进行大胆操作及履行责任所必须具备的设备和部件。但是，依据其所存在的环境，该机器可能发生突然且急剧变化。

如该机器发生故障或小故障（如致盲部件遭受破坏），那么该机器将排气直至停止运行。如您注意到该机器不断出现危险行为，请关闭机器以避免具有自动毁灭性倾向的危险发生。

Antipersonnel Nymphomaniac Wanderer

Instructions:

Rent this machine to brave the danger and bring you back to flat out rotten species, decomposed biomass, from any 'no-man's land.

The Wanderer can be transformed for collecting other materiel. All robot 'tuning' of terminations, articulated arms, legs and tips is authorized, under the condition that you return the machine in its initial state.

The machine collects any ingredients to be recycled for new productive uses. This grants a second life to the waste, and the trash in polluted areas such as post-military zones with unreachable infrastructure interstices.

Legends and fairytales are simultaneously transported out of the depth of those abandoned situations, as in a 'Stalker' experiment to touch the unknown. Please take care of the backlash of those creatures.

Its first use and development was for the 'itshootmedown' experiment.

Precautions for use:

The machine is originally built with a very high self-estimation sensorial device, as well as a danger-blinding component, both necessary for its brave actions and responsibilities. However, depending on the environment it is exposed to, the machine could be subject to sudden and violent changes in self-esteem.

In the case of failure or minor breakdowns (such as the danger-blind component becoming damaged), the machine will exhaust itself to the point of self-destruction. If you notice that the machine keeps exposing you to dangerous situations, switch it off to avoid risks of suicidal tendencies disguised as bravery.

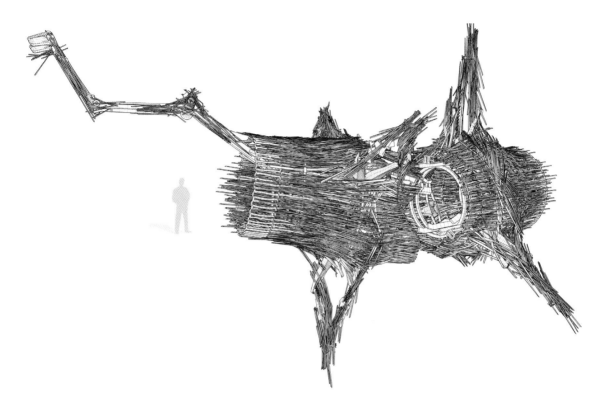

内向型回收器

使用说明：

该机器回收冶金及建筑工地上的垃圾，并通过将凌乱的堆砌物转换成畸形堆砌物的方式将其改造成潜在穴居形态。

目前，该机器仍在研发中。因此，可以优惠的价格租赁该机器进行贝塔测试。为有效地收集钢材所需的能耗级至今还未能准确的测定。该机器在运行时极易发生故障。我们建议仅允许该机器与起重机一起租赁，这是因为起重机有助于稳定该机器的程序和定位。我们要求客户提供反馈意见以便优化设计。当前设计的机器看似比较柔弱。如果没有进一步的让它存在理由，那么将从目录中去除该机器。

注意事项：

由于该机器解析度欠佳，极易受伤害。

需要通过赋予该机器个性特征来保护其免于丧失身份感，否则，其可能混淆本身与建成环境，并形成伪装表皮将自己隐匿于其本身的构造之中。在逐渐丧失个性的早期阶段，该机器可能被其形体金属化趋势所侵蚀。

Introverted Eczemetal Recycler

Instructions:

Transforming informal heaps into deformed ones, this machine recycles waste from metallurgic and construction sites into potential troglodyte morphologies.

This machine is still in development. It is thus available for rent for a special discount as a beta test. The provision of sufficient energy levels for the effective gathering of steel has not yet been gauged accurately, and dysfunctions may easily occur when the Recycler is in operation. We recommend that this machine only to be rented in parallel with the crane that is able to stabilize its agenda and positioning. We require feedback from customers to improve the logic behind its design, which appears for now weak. This machine will be removed from the catalogue if there are no further reasons for it being in existence.

Precautions for use:

Due to its lack of resolution, this machine is especially vulnerable.

Protect it from the feeling of identity loss by engaging with it on a private level – otherwise it might show a tendency to confuse its own being with the built environment, and develop skin camouflage diseases in order to disappear inside its own construction. An early stage of depersonalization can be spotted by its tendency towards metallic somatization.

身体建设机器

使用说明：

出租的这台超级蛋白质装置十分灵敏，能够通过挤压阴茎状的竖向管子排出液体混凝土，然后将混凝土转变为固体凝结物，之后能够抵抗地心引力的作用并继续加工工作。

该机器出租下限至少为30个家庭，这是根据生命政治学做出的决策。

该机器是能够操作和使用的机器，可以用于自治的微城市中自下而上的系统。30多个家庭被称作是"群体"，能够驱动他们自己的构造系统熵和"集体生活"系统。系统设计依据了当代生命科学的潜力以及对人类生理机能和化学机能的平衡的重新解读，能够将"动物性身体"（无头身体）所产生的情绪、身体的化学反应，以及个体面对特定情形和环境时的冲突所产生的适应性、同情心和移情直观化。该施工流程通过"机器主义"研发，即行为不可预测，创造了秘密的编织机器，能够利用各种原材料（生物塑性水泥）通过挤压和烧结（全尺寸3D打印）形成化学聚合物，并构成依据计算轨迹的物理形式。这种不规则类书法结构的过程的工作原理类似于机械工的切割术，通过永久性地制作不规则形状来产生一系列连续的几何图形，无需标准化，无需重复，这个技术怪才的涌现仅在于其所依据的流程和协议。

该机器首次使用时用于"anarchitecture_deshumeurs"试验。

注意事项：

机器处幼年阶段与成年阶段之间，为充分发挥机器的工作能力，这两个阶段内的很多问题都还未处理，如成年阶段的排出行为和裸露癖，都是机器的常见症状，应当视为健康的象征。

如机器处在公共社交十分活跃的区域，这些行为可能会继续发展成为性反常行为，表现为症状加剧，变形，超出正常范畴的可怕的性表达等。强烈建议不要将机器放置在公共场所（即您自身所处群体之外的地方）。

该装置同时也有轻度自恋症，导致在具有相似装置的不同群体中会产生强烈反应。

Body-Builder-Shitter

Instructions:

Rent an agile hyper-protein device, shitting liquid concrete through a vertical phallic extrusion, which is turned into coagulations that it stands on to continue the construction process in defiance of gravity.

The Shitter is only made available to rent to a minimum of 30 families, dedicated and driven by bio-political decisions.

The device is a usable, operative machine for a self-organized micro-urbanism conditioned by a bottom-up system. The 30-plus families, called 'the multitude', are able to drive the entropy of their own system of construction, their own system of 'vivre ensemble'. It is based on the potential offered by contemporary bioscience, the rereading of human corporality in terms of physiology and chemical balance to make palpable and perceptible the emotional transactions of the 'animal body', the headless body, the body's chemistry, and information about individuals' adaptation, sympathy, empathy and conflict when confronted with a particular situation and environment. The construction process developed through 'machinism' – indeterminate and unpredictable behavior – with the creation of a secretive and weaving machine that can generate a vertical structure by means of extrusion and sintering (full-size 3D printing) using a hybrid raw material (a bio-plastic cement) that chemically agglomerates to physically constitute the computational trajectories. This structural calligraphy works like a machinist stereotomy composed of successive geometrics according to a strategy of permanent production of anomalies: with no standardization, no repetition, except for the procedures and protocols at the base of this technoid slum's emergence.

Its first use and development was for the 'anarchitecture_deshumeurs' experiment.

Precautions for use:

The machine is set in between anal and foecal stages, leaving both unresolved in order to achieve full development of its construction capacities. Anal expulsive behaviors, as well as exhibitionism, are frequently displayed by the machine and are to be considered as signs of good health.

Placed in an extremely social zone, these behaviors could later develop into paraphilia – manifesting in hyperbolic intensifications, distortions, monstrous fruits of erotic expression beyond normal eroticism. It is strongly recommended, therefore, not to place it in public zones (i.e. outside of your own multitude).

The device is also slightly narcissistic, which could provoke strong reactions in similar devices of different multitudes.

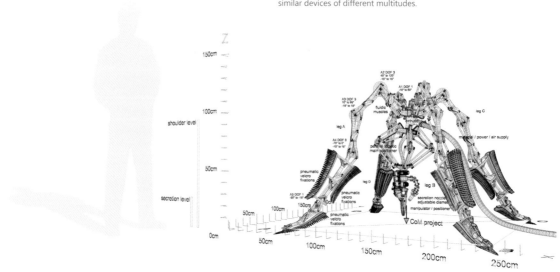

差异与重复 / 复杂的随机装置

使用说明：

出租的这台机器可以沿着机器运动的轨迹填充曲面形态。该机器拥有数个机器臂，可以进行强制性的关节运动，并可以进行复杂的舞蹈动作，能够产生理想的可编程经验外形和结果。

该机器连同特定数量的部件共同出租（仅限于500个零件的组件），以便在任何情况和条件下使用。单个部件采用尼龙搭扣，通过梳齿状设计自动扣紧，位置可变，能够方便地产生多种结构，无论是巨大物体、液体、不透明还是透明物体。

机器安装后的尺寸能够达到3m×2m×1m，包括长10m的轨道。请参阅安装说明，以便了解机器/部件在地面上的两种安装位置。你将接受有关逆序电影流程方面的培训，接受培训后能够首先手绘结构和空中曲线，掌握机器窍门，然后探讨你的手部运动轨迹如何成为部件堆叠的轨迹（上限为4m）。精密的制作过程将遵循你在反复改进后确定的曲线。

注意事项：

由于机器的不可预测性，可能会出现极端紊乱，间歇性躁狂，以及不同程度的轻度躁狂和抑郁症。尽管这些症状都是随机复杂过程中不可避免的，长期如此却可能导致副作用，如思绪翻腾和休息不足（关闭模式）。

尽量不要让机器工作过久，避免机器在无休止的繁琐劳作中感到过度疲劳。

相反，如机器反复不断、有顺序地或有系统地出现上述症状，请立刻将其送回商店，以便紧急重新编程。

Difference and Repetition / Intricate Randomizer

Instructions:

Rent this device to populate a surface that will be revealed by the trajectory you cover in response to impulses from the machine. Its multiple arms will follow a dance of intricacy in compulsive, articulated movements, giving ideal programmable empirical shapes and outcomes.

This machine has to be rented with a specific number of components (only available in packages of 500 units) to be populated in any condition, any situation. Each individual component is developed as a Velcro strip, that can be attached using a comb-feather design, with variable positions able to assume – at your convenience – polyphonic structures – be they massive, fluid, opaque or transparent.

The machine is able to be packed in a container measuring 3 x 2 x 1 meters (10 x 7 x 3 feet) including the tracks that are 10 meters (33 feet) long. Please refer to the installation instructions for ascertaining the dual positions of the machine/component on the ground. You will be trained in the inverse cinematic process that will enable you to draw first the structure manually as curves in space by manipulating the machine tips, and secondly discover how the footprint of your handy movement is becoming the trajectory of the components stacking, automatically repeated and assembled by the machine (4 meters/13 feet high maximum). The intricate packing fabrication will follow the isocurves you defined in the space in a repetitive adaptation.

Precautions for use:

Due to the requirement of unpredictability for its work, the machine is subject to bipolar disorder, alternating manic, hypomanic and depressive episodes of varying lengths. Although these episodes are necessary to the nature of the random intricacy process, they might in the long run cause side effects such as racing thoughts and rest (mode OFF) deprivation.

Take care of possible exhaustion of the machine, as well as of the feeling of impuissance in front of its never-ending operations.

On the contrary, if the machine shows repetitive, ordered or systematic combination processes, bring it back to the shop immediately for emergency reprogramming.

OCD包装机

使用说明：

这个极其高效的包装、下订、分类、计算和三维定位机器可用作无限的堆叠和错开操作。本包装机仅供长期租赁。

通过测试是否存在可能以损毁以前状态，来开发一个带多种不确定轨道的环境迷宫。基于此，本机器可以用来扩展已有的构筑物。其创造的这一形态学上的把戏既是一个监狱，也是一个保护装置。这一双重策略避免了使用人感觉到以前的狂怒，并保护他们以免自身机体病变。

参与人要求签署私人协议并以自愿犯人的形式参加这个游戏，在游戏中，他们被永久地随意包裹。如果必要，你可以随时使用PDA上的RRID，以便重新进行定位，但这是要冒险的。

该机器的首次使用和开发是针对"Olzweg"试验。

注意事项：

为了取得高效率的下订、计算、安排、检查以及清洁等作业，本机器植入了入侵性思维，可能会产生不安、忧虑、恐惧和担心。

这些旨在减少焦虑的重复性行为也会在在对特定数字的厌恶中或者在紧张仪式的大量重复中被放大。

如果你注意到这种强迫性失调信号，请立即将机器带回店内以便减小输入焦虑。

OCD Packer

Instructions:

Rent this extremely efficient packing, ordering, classifying, numerating and xyz-positioning machine, for endless stacking and staggering. The Packer is only available for long-term rent.

The machine works to extend existing construction, by testing the possibilities of wrapping, smearing and invading a previous situation to develop a surrounding maze with multiple uncertain trajectories and 'parcours'. The morphological trap it creates is both a jail and a protection apparatus. This dual strategy avoids the occupants perceiving their own madness and protects others from their own pathologies.

Participants require a personal agreement and discharge to play this game as a 'voluntary prisoner', lost in the permanent entropy of packing. In any case you could use, if necessary, RFIDs on PDAs to rediscover positioning – but at your own risk.

Its first use and development was for the 'Olzweg' experiment.

Precautions for use:

In order to achieve high efficiencies in ordering, numbering, arranging, checking, cleaning, etc, the machine has been implanted with intrusive thoughts that can produce uneasiness, apprehension, fear and worry.

The repetitive behaviors aimed at reducing these anxieties can also manifest themselves in an aversion to particular numbers or in the absurd repetition of nervous rituals.

In case you notice such signs of obsessive compulsive disorder, please bring the machine back to the shop immediately for a reduction of input anxieties.

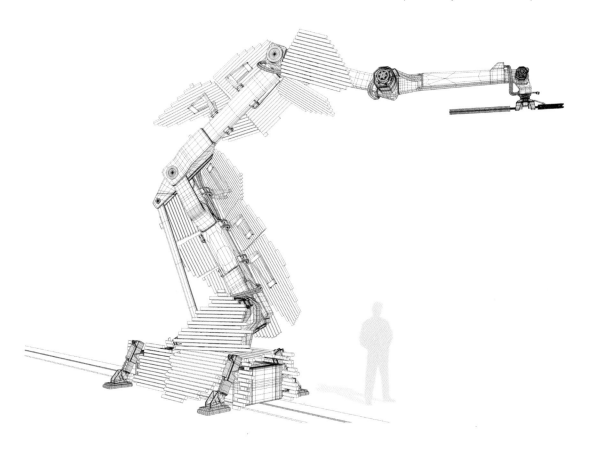

藻类萨克循环器

使用说明：

该水下装置为可清除海水中海藻、提取化学物（方解石）及微粒的提取器。逐渐形成的堆积物受到涌流及潮汐的不断冲刷，涌流及潮汐左右了结晶的方向及进程，位置不可预测。

本机器仅可在海水中使用（海水每升含钙约400mg，1.6t/m³）。钙来自诸如石灰岩、大理岩、方解石、白云石、石膏、氟石及磷灰石之类的岩石溶解。在租用该机器前，您需要进行调查，以确定当地的钙量。我们可以提供专业技术。

为了能够正常运行，水深应在6~20m之间。

提取及转化过程为专利技术。化学过滤及化学反应不在本使用说明中透露。请勿打开机器的密封核心部件，因为有毒性。

注意事项：

本机器受外部因素影响，如潮流、潮汐及月球偏心率。受水冲刷越严重，越能良好运行。本机器在不良环境下，也会出现故障。

考虑到这些综合特征，本装置可能变为完全不可使用的结构，但不能将责任归咎于该结构本身。租用本机器时，用户需自己承担风险。

在极端恶劣的情况下，如机器过度暴露在水中或其他环境因素条件下，可能会出现故障。最终，变得完全不可使用。

Algae-Sacher-Cyclothymia

Instructions:

Rent this under-seawater device that behaves as an extractor removing algae and extracting chemicals (calcite) and particles from the water in order to agglomerate a masochism structure. The progressive accumulation is condemned to be pulled and pushed by the current and tide, which drives the orientation and the progression of the crystallization without a forecast positioning agenda.

The machine is usable only in seawater, which contains approximately 400 milligrams per liter of calcium and represents 1.6 tons per cubic kilometer. The calcium is obtained from dissolving rocks such as limestone, marble, calcite, dolomite, gypsum, fluorite and apatite. Before renting you need to request a survey to confirm the quantity of calcium in your location. We can provide this expertise.

In order to function, the device requires a water depth of between 6 meters and 20 meters (20 and 65 feet).

The extraction, transformation processes are patented. The chemistry filtering and reaction cannot be divulged in these instructions for use. Please do not open the sealed core of the machine; it is toxic.

Precautions for use:

The machine is built with a total submission to external factors such as currents, tides and lunar eccentricities. The more it is ill-treated by the water, the better it will work. The machine is also cyclothymic, subject to mood swings, and is volatile in its responses to the water humiliations.

Due to the mixture of these characteristics, the device is susceptible to construct totally useless structures, and cannot be held responsible for the unusable nature of the structures built. You are renting it at your own risk.

In extreme cases of maltreatment, where the machine is overexposed to water or other environmental factors, it could become self-defeating or suicidal. Ultimately, it could completely stop functioning.

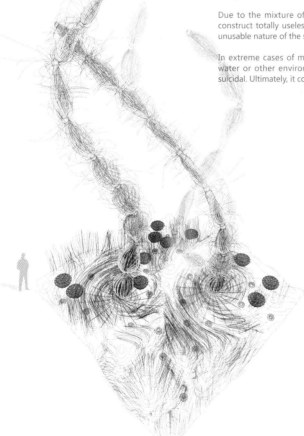

贪食症外壳-编织机

使用说明：

这台蚕茧编织机器十分精密准确，可打造出临时性建筑、野营地、户外工坊或露天派对。

不要抱怨这个机器即是产品的制造者又是产品结构部分，把自己困在了自己编织的网中。它是自己技术流程的实施者。

丝膜可以防水，也可以不防水。请参见机器说明中的编织密度。网线由生物产品、淀粉及亚麻制成。其生命期限大约为10天，此后则将降低并失去结构阻力。融解条件为百分百天然，降解过程还将为地面提供氮及营养元素。无需担心会产生表面污染。

可根据要求，提供不同生命期限的人造丝。

人造丝由10km（6mi）的线筒提供。

注意事项：

可调整机器，使其表面缺少对实体的感知，从而不间断地编织表面。但机器可能会无法预料地排斥表面，返回自身结构，以减少自身四周防护茧的形成。

贪食症趋势将根据潜在特质重新打造出虚拟维度，自身的连接、情绪及运动可能产生未来其产品内部的缺陷。在此阶段，无法停止幽闭恐怖症这一过程。

Bulimic Enclosure-Weaver

Instructions:

Rent this silk cocoon-weaving device – preciously precise and accurate – to create temporary buildings, camping sites, outdoor workshops or garden parties.

Do not complain that this machine is both the producer and the structure of the production, trapped in its own net. It has its own process of know-how.

The silk membrane could be made waterproof or not. Please refer to the density of knitting in the machine's instructions. The wire is the product of bio-production, starch and flax. Its lifespan is around 10 days before it degrades and loses its structural resistance. This melting condition is 100 per cent natural, and the process of necrosis will provide nitrogen and nutritional elements to the ground. Do not be afraid of the ostensible pollution it seems to generate.

Different time spans for synthetic silks are available on application.
The synthetic silk wire is provided by a bobbin of 10 kilometers (6 miles).

Precautions for use:

The machine is conditioned to have a lack of bodily feeling in its surfaces in order to keep it endlessly weaving surfaces. However, it can unexpectedly reject the surface and return to its body, inducing the formation of a protective cocoon around itself.

This bulimic tendency to recreate a virtual dimension of potential traits, connections, affects, and movements around its own body are symptoms of the future loss of the machine inside its own production.

At this stage nothing should be attempted to stop the claustrophobic process.

星盘预言机

使用说明：

出租的这台星盘预言机可以确定两个独立星球-太阳及月亮的威胁级别，以及它们导致的人体病理学。它可以探测这些星球带来的潜在危害，确保与天穹的沟通：保护您免受太阳磁暴及辐射的危害，以及月亮带来的心理性创伤的折磨。

除了日（月）（每隔6,585.32天；准确地说，为18年10或11天8小时，根据闰年的时间确定）以外，星盘预言机不能将太阳及月亮控制在同一水平位置。

机器的太阳预言部分表明了星球的天体周期。机器重点关注太阳方位上臭氧层缺口，以及臭氧层提供的紫外线防护度情况，臭氧层可以减弱紫外线的影响。粉末铀具有显示紫外线发射强度的自然功能，可利用其与机器协作。由于粉末铀可发射阿尔法射线（位于规定临界值以下，已经取得法律许可），因此需在特定条件下使用粉末铀。机器具有双向偏执性：一种有害物质为另一种有害物质的传感器，带来了一系列关于过去-现在-未来的工业附属效应。

机器的月亮预言部分：引力以及对转变的恐惧（现实或幻觉）。通过一种超自然科学的方法，它发挥了"想象的科学"这一指向的作用。所有一切看起来均不切实际，但所有这些又影响着你的新陈代谢。

该机器首次使用时用于"theBuildingwhichneverdies"试验。

注意事项：

由于物体的双重特性，低强度精神分裂症状的发作是很正常的，甚至对于机器的有效运转来说十分必要。

如将机器置于模糊的坐标位置，钟表受星体运动的迷惑性及偏执的解释影响，将引发混乱的报告及图铜，在正常情况下无法阅读或理解。

幻觉发作的原因可能包含第三个星体，或对荒诞、致命事件的预测，如过度暴露于危险的月亮紫外线中或受到狼人的日间入侵。

在极限情况下，机器将从漫无目的的躁动及运行转向完全的紧张症。如发生这种情况，建议关闭电源。

Algae-Sacher-Cyclothymia

Instructions:

Rent this device – the Astrolabe Stutterer – to ascertain the level of threat posed by two discrete planets, the sun and the moon, and the human pathologies they produce. This machine detects any potential harm that these planets threaten, and secures your negotiation with the celestial vault: protecting you from magnetic storms and radiation from the sun, and the psycho-lupus affliction of the moon.

The Astrolabe Stutterer cannot simultaneously maintain an equal position between the sun and the moon, except during eclipses, every 6 585.32 days, exactly 18 years, 10 or 11 days and 8 hours, depending on the occurrence of leap years.

The part of the machine dedicated to the sun indicates the planet's celestial cycle. It particularly highlights any gap in the sun's position and the degree of protection afforded by the interaction of solar rays with the ozone layer, which has had its impact depleted by UV emissions. The device can be used with uranium powder, which has a natural afterglow that indicates the intensity of UV emissions. The uranium powder is provided with special conditions, because of the emission of alpha rays (below the administrative threshold), which have been agreed by legal settlements. This machine has the potential for a double paranoia: one harmful substance acts as sensor to another, providing a chain of past–present–future industrial collateral effects.

The moon part of the device points to the symptoms of the moon: the forces of attraction, and fear of transformation (real or illusionary). It works as a vector of 'science of the imaginary', through a pataphysical approach. Nothing seems real, but everything in fact affects your metabolism.

Its first use and development was for 'theBuildingwhichneverdies' experiment.

Precautions for use:

Due to the dual nature of the object to be read, schizophrenic episodes of low intensity are normal and even necessary to the effective functioning of the device.

If placed under ambiguous coordinates, the clock is subject to delusive and paranoiac interpretations of the astral movements and this will induce disorganized reports and drawings, impossible to be read or understood under normal circumstances.

Hallucinatory episodes may include the creation of a third aster or the predictions of absurd mortal events such as over-exposure to the moon's dangerous UV light or again a daytime invasion of werewolves.

If pushed to its extreme, the device will run from purposeless agitation and motions to complete catatonia in which case it is recommended that it be unplugged.

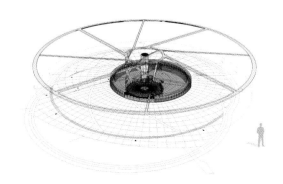

CASE STUDIES
案例分析

Höweler + Yoon Architecture
Höweler + Yoon建筑事务所

Matter Design Studio
Matter设计工作室

The Bartlett School of Architecture, UCL
伦敦大学学院Bartlett建筑学校

ICD Institute for Computational Design, University of Stuttgart
德国斯图加特大学计算设计学院

Harvard Graduate School of Design
哈佛大学设计研究生院

CAUP, Tongji University
同济大学建筑与城市规划学院

Greyshed
Greyshed设计事务所

项目团队

设计：尹美真教授
建筑：Höweler + Yoon 建筑事务所
结构工程：尼佩尔斯·汉尼普斯
砌筑顾问：Ochsendorf DeJong砌块顾问公司
景观结构：Richard Burck事务所
土木工程：Nitsch工程公司

Project Team

Design: Professor J. Meejin Yoon
Architect: Höweler + Yoon Architecture
Structural Engineer: Knippers Helbig- Advanced Engineering
Masonry Consultant: Ochsendorf DeJong and Block Consulting
Engineers Landscape Architect: Richard Burck Associates Civil Engineer: Nitsch Engineering

"MIT肖恩·科利尔纪念碑"项目建成实景（图片版权：伊万·本）
The MIT Sean Collier Memorial (Image Credit: Iwan Baan)

Robotic Stereotomy
The MIT Sean Collier Memorial
机器人切石法
MIT肖恩·科利尔纪念碑的建造

J. Meejin Yoon, Eric Höweler / Höweler + Yoon Architecture
尹美真，埃里克·霍威尔 / Höweler + Yoon建筑事务所

MIT肖恩·科利尔纪念碑位于科利尔军官于2013年4月中枪身亡之地。纪念碑采用当代砌体结构，成为项目基地的标志性建筑，传达了"威武的科利尔"的勇气和团结精神。纪念碑采用五向平板石穹顶，穹顶采用五面细长的径向壁支撑，将当代的数字建造技术、结构分析软件和计算技术同古老的纯受压穹顶技术相融合。石拱是建筑结构组织中最基本的元素，材料在空间中的排列顺序以石拱为基础，同时传递了压力荷载。石拱的几何造型直观地指示了力在材料中的流动，将荷载从一个石块传递到另一个石块，并将力转化为形式。

设计选用不配钢筋的砌块结构来纪念科利尔军官，这具有重大意义。巨大的石拱穹顶较薄，营造了悬浮感和轻盈感，而每个石块的锥形结构则同整个砌块拱顶的几何造型保持了一致。不配钢筋的砌块结构设计要求采用一种新的找形方法，能够使结构的所有各部分在各种荷载情况下保持压缩状态。拉索式的几何造型完全遵循力线，将砌体材料内的推力线视觉化。模拟技术遵循了"胡克定律"原则，界定了在特定荷载下纯受拉垂链和纯受压倒拱之间的一致性。安东尼·高迪曾将加权物理模型运用于砌体的几何找形中，弗雷·奥托则将这种模型运用于慕尼黑的奥林匹克体育馆，这两个都是著名的例子。这些技术采用

Situated where Officer Collier was shot and killed in April of 2013, the memorial marks the site with a contemporary masonry structure—translating the phrase "Collier Strong" into a structure that embodies both strength and unity. Consisting of a five-way flat stone vault buttressed by five slender radial walls, the design for the Sean Collier Memorial combines age-old techniques for spanning vaults in pure compression with cutting-edge digital fabrication technologies, structural analysis software, and computation. The stone arch is among the most elemental of structural organizations, ordering materials in space and transferring loads in pure compression. The geometry of the arch visually indexes the flow of forces through material, transferring loads from stone to stone, and translating force into form.

The choice to use an un-reinforced masonry structure to memorialize Officer Collier is significant. The shallowness of the massive stone vault overhead creates an effect of suspension and weightlessness, while the tapered geometry of the individual stone blocks reveals the keystone geometry of the masonry arch. The design of un-reinforced masonry structures typically requires a form finding method, whereby all parts of the structure

物理模型"计算"拱顶或薄膜的几何造型，检测结构承担自重和活荷载所产生的作用荷载的能力。MIT肖恩·科利尔纪念碑采用了逆序施工法组建，最先安装的是楔石。结构组建及其设置顺序采用串联法，并同每个石块的建造顺序同步。

科利尔纪念碑的设计以过去的模拟技术为基础，代表了当代建筑生形的一种综合方法，即利用各种工具、方法、模型及软件平台的组合以实现整体形态。设计优化的核心为力多边形的迭代分析及修正，以及对穹顶每块石材中所有力矢量的测量及重新排列。这一方法（与计算静力学图解共同生成）一般由二维的简单弧结构完成。五向穹顶布局要求在5个独立部分内进行静力学图解计算，将穹顶分为半径饼图分区。由Ochsendorf DeJong砌块顾问公司开发的计算工具为概念设计分析了5个方向的穹顶几何构成，以确保重力荷载下穹顶的整体稳定性。作为Rhino的一个插件程序，RhinoVAULT主要用于生成外壳结构的膜的解决方案，它可以探索并计算石材内的三维压缩解决方案。

将整体穹顶造型细分为楔形块时，必须提供详细节点样式的设计，以保持纪念碑的整体平衡。石材间的接缝线与自重条件下的力轨迹线垂直，而凹口的节点样式将使石材的滑动风险降到最低，以确保每块石材都在周围石块的合力作用下得到牢固支撑。该过程的最后一步是利用计算脚本来核实最大的石材的尺寸位于单个石块砌筑技术的尺寸范围之内。

remain in compression during all loading scenarios. Funicular geometries explicitly follow the lines of force rendering visible the thrust lines within the masonry material. Analog techniques utilize the principle known as Hooke's Law, which defines the correspondence between a hanging chain in pure tension and the inverted arch in pure compression for a given set of loads. Weighted physical models were famously used by Antoni Gaudí to form-find geometries in masonry and by Frei Otto to design the Olympic Stadium in Munich. These techniques use a physical model to "compute" the geometry of the arch or the membrane, and test the ability of the structure to support the range of applied loadings produced by self-weight and live loading. The Sean Collier Memorial is assembled using a historically reverse construction technique of centering, in which placement of the keystone initiates the erection sequence. Assembly of the structure and its setting sequence works in tandem with the fabrication of each stone piece.

Building on the analog techniques of the past, the design of the Collier Memorial represents a synthetic approach to contemporary form-making, using a combination of tools, methods, models, and software platforms to arrive at an overall form. Central to the refinement and optimization of the design is the iterative analysis and modification of the force polygon, mapping and reconfiguring all the force vectors in each stone of the vault. This method, generated with computational graphic statics, is typically performed on simple arch configurations in two-dimensions. The 5-way vault arrangement requires the graphic statics calculations to be performed in five individual segments, dividing the vault into radial pie-slices. Computational tools developed by Ochsendorf DeJong and Block analyzes the five-way vault geometry for conceptual design to ensure the global stability of the vault under gravity loads. RhinoVAULT, a plug-in for Rhino that generates membrane solutions for shell structures, explores and computes three-dimensional compressive solutions within the stone.

The subdivision of the overall vaulted figure into voussoir blocks requires the design of a jointing pattern that contributes to the overall equilibrium of the memorial. The joint lines between stones are designed to be normal to force trajectories under self weight, while the pattern of notched connections minimizes the risk of sliding between stones and ensures that each stone is held securely by the resultant forces of the adjacent blocks. The last step in the process uses computational scripts to verify that the largest stones are within the bounding box limits of quarrying technology for a single block.

Fabrication of the Sean Collier Memorial is a combination of traditional hand methods and digital manufacturing. Manufacturing each stone block relies on the combined efforts of robot and mason. Quarried stone is first cut into parallel-faced slabs by a

现场组装
Assembly on Site

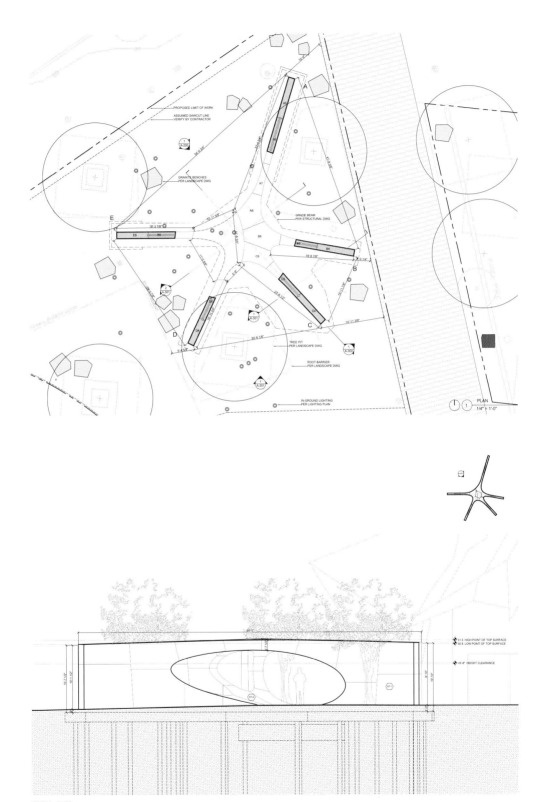

平面与立面
Plan and Elevation

单轴机器人石锯首先将开采的石材切割成平行的石材板
Quarried stone is first cut into parallel-faced slabs by a single-axis robotic block saw

肖恩·科利尔纪念碑的建造是传统手工方法与数字切割的完美结合。石块的制作依靠的是机器人与泥瓦匠的紧密协作。开采的石材首先由单轴机器人石锯（锯片长3.5m，重达81kg）切割成平行的石材板。每个方向的切割都要求砌筑工人对石砌块进行物理再定位。一旦露出石材的外部粗糙几何结构，泥瓦匠将对石块进行检查，并移至固定桥数控铣床或KUKA 500机器人处。体积相对较小的石块可放入数控机器的作业空间，而较大的石块必须由KUKA 500机器人在转台上进行切割，机器人接头可承受544kg的压力，并以2.5m/s的速度运送石块。为满足个别石块的复杂几何结构要求，这些机器可连续工作几个星期。最早批次的石块可一天24h连续制造14天。

设计依赖于33块石块的精确配合，以整合成一个整体造型。为确保嵌入在多个石块中央空间的视觉连续性，各个接头的尺寸都应控制在最小，即6.35mm，这就要求格外精准的建造。机器人切割开始前，泥瓦匠必须登记具体石材编号，并确定石材位置，使之符合数字Rhino模型的要求。一旦将石材位置传达至机器人，机器人将进行一系列的铣削或锯切步骤，切割至完成建模表面的2~3mm。机器人铣削过程最终生成石块的实际容差在0.5mm以内（实际石块与数字模型的容差）。石材制造过程中如此高精度的要求带来了独特的挑战，同时，这也证明了制造过程中对石材的每次切割及每次测量的精度。即使是复杂的3D扫描仪，也很难把握0.5mm的容差。此外，随着切割

single-axis robotic block saw, which has a 3.5 meter blade that weighs 81 kilograms. Each directional cut requires a physical reorientation of the stone block by masonry workers. Once the rough, exterior geometry of the stone is revealed, the piece is examined by masons and moved to either a fixed-bridge CNC milling machine or a KUKA 500 robot. Relatively smaller pieces fit into the work envelope of the CNC, but larger pieces must be cut on a rotating table by the KUKA 500 robot whose joints can accommodate a load of 544 kilograms and move the piece at a rate of 2.5 meters per second. To accommodate the geometric complexity of particular pieces, these machines are able to constantly cut for weeks at a time. The first few blocks were fabricated continuously 24 hours a day for 14 days.

The design relies on the exact fit of the 33 stone blocks to coalesce into a singular form. To ensure the visual continuity of the central void across multiple stones, joints were constructed at the minimum 6.35mm, which requires extreme precision in its fabrication. Before robotic cutting begins, masons must register and orient the actual stone to correspond with the digital Rhino model. Once the location of the stone is communicated to the robot, a series of milling or sawing steps removes material to within 2-3 millimeters of the finished, modeled piece. The robotic milling process produces final stone pieces that are within a 0.5

KUKA 500机器人在转台上对大块石材进行切割
Larger pieces must be cut on a rotating table by the KUKA 500 robot

的进行,石材的材料性能也逐渐变弱。尽管锯片以精确的直径开始切割,当锯片与石材接触后,锯片也会受到磨损。在生产过程中,需一直修正石材及工具。建造要求对锯片进行连续的校准,以确保机器人切削工具的切割准度。

肖恩·科利尔纪念碑提出了"局部到整体"的问题,它是一个由多个部分按照几何尺寸构成的单一整体结构。穹顶设计体现了材料结构中的多个结构性原则,以使索状结构成为可能。这种将力可视化的做法与麻省理工学院开放、透明的理念相一致,而需要五面墙才能形成稳定形态的想法也象征着整个社区对科利尔军官的共同怀念。

从方法论上来说,肖恩·科利尔纪念碑的设计过程涉及实体模型的建设(泡沫、木材、石材及3D印刷粉末)与数字工具的模拟间的一个来回过程。实体模型可以对设计进行机械测试,同时能够预演其安装程序。利用实体模型,可事先进行震动模拟及稳定性试验。数字模型可用于分析、优化设计,以较少材料消耗,并确保设计的合规性。数字工具的开发产生了古老建筑结构技术的现代形式,这强调了方法论的多样性,当代设计并未赋予任何单一的设计工具任何特权,它受益于数字技术及材料计算技术两个方面。利用计算工具及高度精确的机器人建造将力转化为形式的方法,凸显了材料性能及设计师对其几何操作能力间的对等性。

millimeter tolerance between the actual stone and the digital model. The need to achieve such a high tolerance during stone fabrication presents unique challenges, such simultaneously verifying the precision of each cut and measuring the physical stone during fabrication. Even a sophisticated 3D scanner has difficulty measuring a tolerance of 0.5 millimeter. Additionally, the material properties of the stone wear away the equipment as it is cut. Though the saw blade begins a cut as a precise diameter, the blade wears away as it comes in contact with the stone. The stone and the tool are being modified during the manufacturing process. The fabrication requires a constant re-calibration of the blade to reestablish the robot with the extent of its cutting tool.

The Sean Collier Memorial raises "part to whole" issues as a structure composed of discreet parts that conform to geometries to form a seemingly singular whole. The vaulted design embodies structural principles in its material configuration to make the funicular structure legible. This didactic visualization of forces is consistent with MIT's ethos of openness and transparency, while the idea that all five walls are needed to achieve a stable form is symbolic of a community coalescing to commemorate a loss.

Methodologically, the design process for the Sean Collier Memorial involved a back and forth process between the construction of

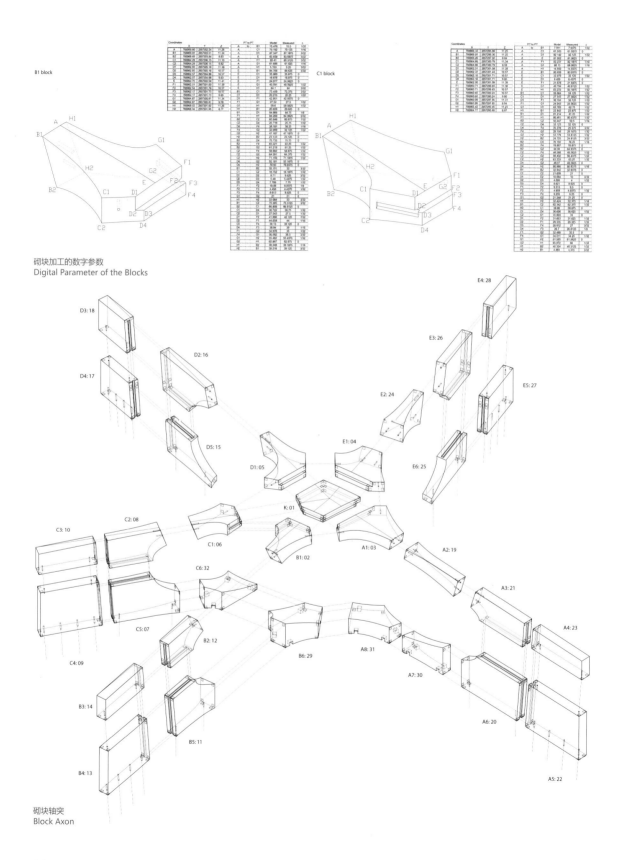

砌块加工的数字参数
Digital Parameter of the Blocks

砌块轴突
Block Axon

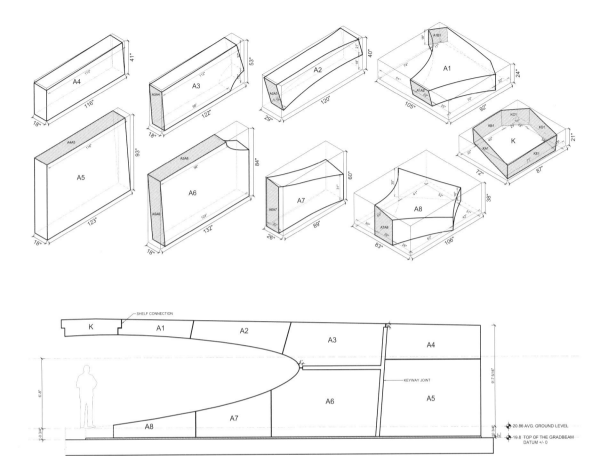

立面及其对应的砌块编号尺寸
Elevation and Its Block Axon

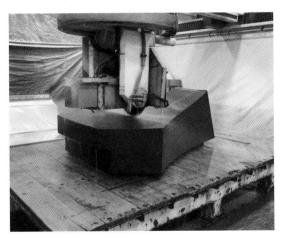

physical models (foam, wood, stone and 3D printed powder), and simulations with digital tools. Physical models allowed the design to be tested mechanically and the erection sequence to be rehearsed. Rocking simulations and stability tests were performed with the physical model. Digital models allowed the design to be optimized and analyzed for material conservation and code compliance. The development of digital tools to produce contemporary forms with archaic building techniques highlights the methodological diversity that characterizes contemporary design which does not privilege any single design tool, but rather benefits from both digital and material computation techniques. The correspondence between the material properties and the designer's ability to manipulate them geometrically is highlighted by the methods of translating force into form with computational tools and high precision robotic fabrication.

2012年，在Banvard画廊展出的"LeFevre拱"项目
Matter Design, La Voûte de LeFevre, Banvard Gallery, 2012

La Voûte de LeFevre
A Variable-Volume Compression-Only Vault
LeFevre拱
一个可变体量的仅受压拱

Brandon Clifford, Wes McGee / Matter Design Studio
布莱登•克利福德，威尔斯•麦克盖 / Matter设计工作室

质点弹簧系统通常被用于发展受压找形系统。本文提出使用质点弹簧系统去响应理想形式，从而生成由体积材料加工而成的可变体量仅受压结构。通过改变系统的深度和体量，通过材料深度被重新定向，生成与结构意义上的最佳形式（均匀厚度）相反的理想形式。本文阐述了如何生成、建造和测试由可变体量单元组成的仅受压拱。这一研究将推动物理体积计算及体积建造方法的发展。
Particle-spring systems are commonly used to develop compression-only form finding systems. This paper proposes to use a particle-spring system in response to a desired form in order to generate a variable-volume, compression-only structure fabricated of volumetric material. By varying the depth and the volume of the system, loads can be re-directed through the depth of material in order to result in a desired form, as opposed to a structurally optimal form that assumes a uniform thickness approach. This paper proposes to generate, build, and test a compression only vault composed of variable-volume units. This research will progress knowledge surrounding volumetric physics calculations as well as volumetric fabrication methodologies.

薄壳型仅受压结构系统对建成环境来说是相对陌生的；而从另一方面看，仅受压结构又是古老的。薄壳结构设想出一种最小而一致的截面，这一设想受到材料效能的驱动，表现为完全受结构因素（通常是重力）影响发展出的形式，因此，"找形"一词应运而生。建筑必须对结构问题做出响应，但同时也不得不解决一系列其他问题，如声学、形式、功能等等，因此，形式不一定全然受结构要求的驱动。例如，哥特式大教堂的石材，其横截面在不同深度均存在推力向量。这些大教堂并非依据理想的悬链线确定形式，而是在双重考虑了仅受压原理和建筑需求的情况下形成的。受这一方法的启发，本文探讨了通过由可变体量构成的仅受压系统生成理想形式的潜力。

Thin-shell compression-only structural systems are relatively new to the built environment. Compression-only structures on the other hand are ancient. Thin-shell structures assume a minimal and consistent cross-section. This assumption is driven by material efficiency. The results are forms developed exclusively by structural concerns (typically gravity), hence the term form-found. Architecture has to respond to structural concerns, but it also has to address a variety of other issues—acoustical, formal, programmatic, etc. It is not necessary for form to be driven strictly by structural requirements. For example, Gothic Cathedrals contain the thrust-vector within the variable depth of

我们已对现有的可变深度结构进行了很多研究，分析在材料深度中是否存在着推力向量，[1]其他研究方法假设采用固定深度材料来生成设计。本文所采用的研究方法是构想出一个理想形式，并允许可变体量对推力向量进行重新定向，从而生成基于结构和形式的可用性设计。如果说通常的研究追求的是结构的轻薄，那么本文追求的则是结构的形式，通过结构来解决建筑问题。本文并不是提倡复古，而是运用先进的途径与方法将那些早已被遗忘的知识重新植入建筑语境之中。

系统、结构和形式响应

质点弹簧系统以连接到线性弹簧的集中质量（质点）为基础。本研究所使用的解算器是西蒙·格林伍德实现的质点弹簧系统的一部分[2]。"系统中的每个质点都有其位置、速度、可变质量、以及概括所有作用力的一个总向量。"[3]龙格-库塔解算器并不需要用于生成悬链线或是荷载分布，但需要用于评估不规则分布荷载。研究假定生成的几何形式并不是理想的悬链线，因此这一方法将多用于分析不规则的荷载分布。基于对弹簧质点系统的探索，前人已经创造出一些虚拟找形方法，如阿克塞尔·克利安的CADenary软件[4]。

只要推力向量存在于系统横截面中段三分之一范围内，仅受压结构就能够站立。尽管我们并不总能预知一个结构是否会倒下，但我们能够预测其是否能够站立。有一篇论文[5]介绍了砌体结构的安全定理，安全定理表明：只要能够在结构的截面范围内能够找到一组处于平衡状态的压力网络，那么仅受压结构就能保持站立。这一方案有可能是下限解，因为当评估现有结构时，我们并不能总是获知力网的确切位置[6]。而本文所采用的方法则能够计算并确保在材料厚度范围内存在着推力向量①。

弗雷·奥托[7]和安东尼·高迪等研究学者制作的找形模拟模型，或是阿克塞尔·克利安开发的虚拟找形工具CADenary[8]，都证明了要控制和预测最终找到的形式结果是有难度的。此外，如果找到的形式和模型之外的作用力不相匹配，我们也很难将两者结合为一个方案。因此，本文提出了形式响应策略。在一端输入理想形式，生成可变体量的解决方案，以便外部作用力和基于解算器的模型之间能够发生交互作用。

方法论

LeFevre拱是由基于解算器的模型计算所得，模型从不完美几何体中生成仅受压结构，它需要一个确定的几何体作为输入端，然后在几何体上开口以改变每一单元的质量。这一动态系统依据体量计算重新配置了单元质量：假如单元A的体量为单元B的两倍，那么其质量相应为单元B的两倍。项目材料必须是连贯且实心的（空心材料不能运作）。正是由于项目的质量和体量分布，使得系统中永远保持着零张力，因此这一计算方法产生了能够"永远"站立的结构。

the stone's cross-section. These Cathedrals are not determined by idealized catenary form, but through a confluence of architectural desires with compression-only principles. With this approach as inspiration, this paper addresses the potentials of compression-only systems to be resolved through a variable-volume in order to obtain a desired form.

Much research has been done in analyzing existing variable-depth structures to determine if a thrust vector falls inside the depth of material [1]. Other methods assume a fixed depth of material in order to generate a design. The method proposed in this paper assumes a desired geometry and allows for a variable-volume to re-direct the thrust vector as a means to produce a viable design that concerns both structure and other formal concerns. If typically one assumes thin, this paper assumes form.

This method is dedicated to addressing architectural concerns with structural results. This paper does not advocate for the reversion to a past architecture. It promotes the insertion of lost knowledge into our current means and methods of making.

Systems, Structures and Form Responding

Particle-spring systems are based on lumped masses, called particles, which are connected to linear elastic springs. The solver used for this research is part of a particle-spring system implemented by Simon Greenwold [2]. 'Each particle in the system has a position, a velocity, and a variable mass, as well as a summarized vector for all of the forces acting on it.' [3] This Runge-Kutta solver is not necessary to generate a catenary (even load distribution), but it is necessary when evaluating an irregular load case. The method applied in this research will always be an irregular load case because it is assumed the resulting geometry is not an idealized catenary form. Particle-spring systems have been explored to create virtual form-finding methods such as Kilian's CADenary tool [4].

A compression-only structure will stand as long as the thrust vector of the system falls within the middle third of its cross section. It is not always predictable that a structure will fail, though it is possible to know if it will stand. A paper [5] introduced the safe theorem for masonry structures. This theorem states that a compression-only structure can stand so long as one network of compression forces can be found in equilibrium within the section of the structure. This solution is a possible lower-bound solution. When evaluating existing structures, it is not always possible to understand where exactly this force network is [6]. The method applied in this paper can calculate and assure a thrust vector falls within the thickness of material ①.

LeFevre拱的柱体细节
Column Detail, Matter Design, La Voûte de LeFevre Banvard Gallery, 2012

---

① 针对无筋砌体结构的下限分析的更多资料，请参见Heyman 1982、 Huera 2001及 Huera 2004。
  For further reading on lower-bound analysis of unreinforced masonry structures, see Heyman 1982, Huera 2001, and Huera 2004.

基本几何体

本文假设基本几何体是由模型的外在因素（声学、形式、建筑规范等）预先确定的几何体。未来的研究将关注如何使结构需求和其他形式的驱动因素之间的关系更加流动和交互。尽管几何形式并不一定严格地从结构需求出发，但形式与结构之间的关系必须相近才能生成理想的解决方案。在之前的计算测试中 [9]，以大多数任意几何体作为输入端，计算都能顺利运作；而体量可变的计算则更为微妙，它需要将一系列的数值输入到系统中去，包括上下边界面。这些边界面将单元深度参数化，使其在形式生成的过程中是可变的；而在可变体量的计算中，则是固定的；同时，可变体量的计算要求各单元之间的节点都应位于系统之内。这些质点均匀地分布在上下边界面之间的基本几何体上，引入质点弹簧系统，在表面进行点的定位和分布——越接近几何体的上部，点与点之间的距离越大。

质点弹簧系统

质点弹簧系统由大量质点、连接质点的弹簧以及质点上不断产生反馈给系统的作用力组成。尽管组成方式保持不变，但系统已重组出各种方案 [10]。本文采用的是上文提及的均匀分布系统。

垂直距离 VS 体积

分析砌体拱的常用做法是利用静态区块分析将一个拱分解为数个多边形。每一多边形的面积决定了垂直推力向量 [11]。先前的迭代计算通常采用高精度的垂直距离为每一质点赋予新的相对质量，而本文则采用不同于面积或距离的体量来决定单元的质量。目前已有类似研究，通过体量来分析和决定某种结构的可行性，[12]本文则利用体量的可变性来确保方案的可行性。质点的位置决定了虚拟推力网络的位置。为了生成方案，这些质点需要在计算过程中不断地移动，直至达到平衡状态。一系列操作发生在计算的每一次间隔，将计算复杂化，不再仅仅是简单的距离测算。新的质点位置生成了三维的泰森多边形，这一多边形同底部的基本几何面边界相交。在生成插值曲线的地方，曲线间相互交叉产生了点；与此同时，中心点（同样为质点）找到了位于上界面的闭合点，生成了垂直于两点连线的圆，该圆形所处的平面成为计算数控机床操作的边界面，作为有效的建造限制。圆形和曲线经过放样创造出曲面，该曲面在系统中被削减。这些曲面的交叉部分挤出至上一表面的最近点，创造出拱顶中作为离散单元的楔形块 ②,③。每一单元包含一个封闭体量，告知整个系统其相对相邻体量的质量。

Form-finding analog models by such researchers as Otto[7] and Gaudi, or even the virtual versions like Kilian's CADenary[8] have proved it is difficult to control and predict the results of the final found-form. Moreover, if that form does not correspond with a force that is external to the form-finding model, it is difficult to resolve the two into a solution. This paper proposes form-responding as approach. Form-responding takes a desired form as input and produces a variable-volume solution to allow for interaction between these external forces and the solver-based model.

Methodology

The vault is computed with a solver-based model that elicits a compression-only structure, from a structurally non-ideal geometry. The model requires a fixed geometry as input, and opens apertures in order to vary the weight of each unit. This dynamic system re-configures the weight of the units based on a volumetric calculation. If unit A contains twice the volume of unit B, then unit A weights twice as much. It requires that the material of the project be consistent, and solid (hollow does not work). The computed result produces a project that will stand 'forever' as there is zero tension in the system precisely because of the weight and volume of the project, and not in spite of it.

Base Geometry

This paper assumes the base geometry as fixed. The assumption is that this geometry has been pre-determined by a force external to the model — acoustics, formal, building-code, etc. Future research could allow for a more fluid and reciprocal relationship between the structural requirements and these other formal drivers. While this geometry is not strictly aligned with structural concerns, it must be close in order to result in a solution. In previous versions of the calculation [9] almost any geometry as input would work. The Variable-Volume calculation is more nuanced. This calculation requires a number of inputs to the system. It requires both an upper and lower bound surface. These surfaces parameterize the depth of the units as variable during the form generation, but fixed during the variable-volume calculation. The calculation also requires a location for the node of each unit to be located within the system. These particles are evenly distributed across a base geometry that falls between the upper and lower bound surfaces. This distribution employs another particle spring system to locate and distribute the points across the surface increasing in distance from each other as they approach the upper elevations of the geometry.

---

② 楔形块，通常是石材，建造拱门或拱顶时采用的一种楔形构件。
Voussoir: a wedge-shaped element, typically a stone, used in building an arch or vault.

③ 考虑到与建造方法的相互关系，将包裹体量的表面几何设置为直纹曲面。
The surface geometries enclosing this volume are generated with ruled surfaces due to a reciprocal relationship with the method of fabrication.

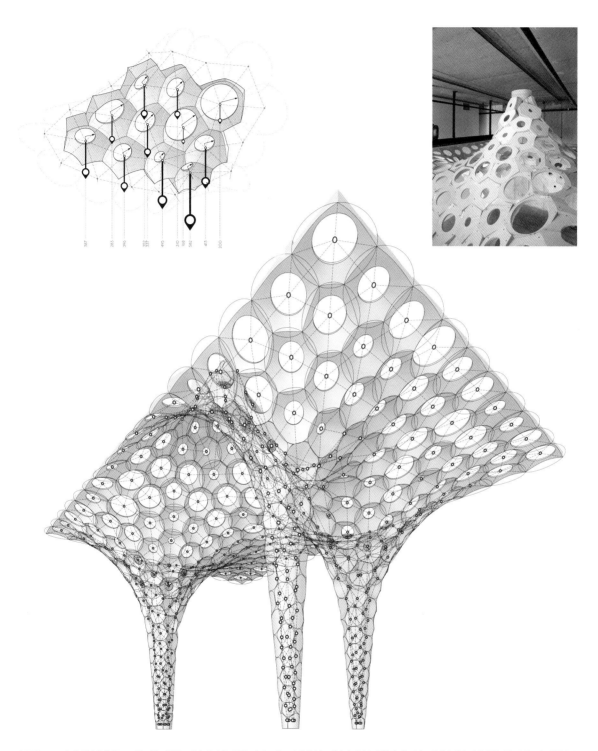

上图说明了可变体量计算结果——位于拱顶的单元随高度升高而逐渐变大，位于立柱的单元随高度降低而逐渐变小。这三类输入信息成为弹簧质点系统重新计算的数据基础。系统将持续进行计算直到找到平衡状态并生成解决方案。
These figures demonstrates the result of the variable-volume calculation—an enlarging of the units in the vault, and a tightening of the units down in the columns. These three inputs serve as the datum with which the particle-spring system computes it-self against. These operations are calculated continually until the system finds equilibrium and a solution can be detected.

等待五轴CNC铣削加工的能组合成理想几何体的粗略聚合构件（上图）。密歇根大学建筑与城市规划学院FABLab工作室的建造支持（下图）。
Roughed aggregated blanks of the desired geometry await the milling operation on the five-axis machine (above). With fabrication support by the University of Michigan Taubman College FABLab (below).

设计

本项目有意实现了从柱到拱的拓扑学转化④，这种转化似乎无任何中断，但事实却并非如此。实际上，柱与拱之间有所差异，柱是被视作单元结构的实心体；而拱则被拆分为多个构成单元⑤。我们试图进行单元间的无缝连接，但木材的纹理暴露了真相。这一虚假的真实世界存在很好的理由去解释：柱并不能扮演拱的角色，柱中的推力向量是垂直的，并非渐渐水平向的。因此，柱不能抵抗水平推力，而能够抵抗纵向弯曲力。柱的主要考虑因素是坚固性。

从实心柱向分散拱的转化，其差异可以通过修饰得以解决。独立单元的修饰延续到柱上，仿佛柱子是独立单元向地面的延伸。这种修饰不是组成拱顶的锥形布尔几何体的简单延续，而是采用了一种新的类似的方法。它参考了锥形布尔几何体，但拒绝对其进行简单的复制。几何的变化使得系统不仅能够调整体量（如拱中的应用），也能够促使从分散到光滑的过渡。随着单元向柱底延伸，它们变得越来越小，但低凹处却慢慢地向柱表面靠近，创造出连续性的错觉，直至将连续体推至基础。这一符号表明，受上方拱顶的重量影响，柱子不得不向外凸出。

Particle-Spring System

The particle-spring system is composed of a number of particles, the length of the springs that connect the particles, and the continual resulting forces on each particle informing the system. While the organization is consistent, the system has been reconfigured in a variety of solutions [10]. This paper employs an evenly distributed system as described above.

Vertical Distance Versus Volume

When analyzing masonry arches, it is common practice to use static block analysis to break down an arch into a few polygons. The area of each polygon determines the vertical thrust vector [11]. Previous iterations of this calculation employed a high resolution of vertical distances to inform each particle with its new relative weight. This paper employs volume as opposed to area or distance. Similar work has been conducted using volume to analyze and determine the viability of a structure [12]. This paper employs the variability of the volume to ensure a solution.

建造

LeFevre拱由波罗的海桦木胶合板制成，胶合板由19.05mm厚的薄板加厚而成。或许这证明了这个行业的现状：获取体积材料十分困难。通过对薄板的数字切割，将每个定制单元切割为这种厚度，之后经过物理重组，生成最终几何形式的粗略体量。将粗略体量编号排在一大块板上，经过胶粘、真空压缩，重新放置于数控机床上。尽管这一流程更费劲，但它比将一整块实心材料直接雕刻成形更加节约材料。

该项目在五轴铣削机上进行，使用的削屑工具致力于花最小的力气切割最多的材料。该路径不使用工具头的末端进行工作，而是使用侧刃铣削 ⑥ 来移除尽可能多的材料。不同于菲利贝尔·德洛姆 ⑦ 的点追踪法，这一方法通过线来追踪几何形态，因此它要求单元由直纹曲面构成。尽管该限制条件放宽了对立柱的要求（一种更为典型的曲面铣削操作能够产生修饰性的凸起），但是其体现了拱顶的锥形布尔几何体为项目的一部分。工具操作的转化同样说明了对立柱及拱顶差异的理解。

分析

本项目采用假设的零填充方式进行建造。因为设计要求拱必须是可拆卸的，因此建造中并不采用砂浆。因为零容差的方法，差异、错误及缝隙不可能得到解决。为确保一些难操作位置的现场施工，团队采用了手工带锯移除问题单元背面的碰撞材料。现场雕刻并不会影响单元的边缘，但却在楔块表面留下了不一致的缝隙，这个缝隙和印加楔体 [13] 流程留下的缝隙恰好相符。泥瓦匠会从墙背面将砂浆填充至石头间的空心楔形中，而墙体前面（建筑表面）则是无砂浆的。关于印加楔体的利用潜力，未来仍有很大的探索空间。

The location of the particles defines the virtual thrust network. In order to ensure a solution, these particles are requiring to be moving during the calculation until they find equilibrium. At each interval of the calculation, a number of operations occur complicating the calculation beyond a simple distance measurement. The new location of each particle generates a three-dimensional voronoi calculation that intersects with the lower bound base geometry surface. This intersection then produces points at the intersection of each curve where an interpolated curve is generated. Simultaneously the centroid point (also the particle) finds the closes points on the upper bound surface and generates a circle perpendicular to the line connecting these two points. The plane this circle is generated on also serves as the flat backside that sits on the table of the computer numerically controlled (CNC) router, a useful fabrication constraint. The circle and the curve are then lofted with each other producing surface that is trimmed with the rest of the surfaces in the system. The intersection of these surfaces extrude to the closest position on the upper surface producing the voussoir② that discretizes each unit in the vault ③. Each unit now contains an enclosed volume that can inform the system with its weight relative to its neighbors.

Design

A deliberate attempt was made in this project to topologically ④ transition from column to vault. No break is inserted in this transition; however, this is a lie. In reality there is a difference between column and vault. The column is solid. It is treated as a single unit. The vault on the other-hand is discretized into its constituent units ⑤. This moment of discrepancy is attempted to be seamless; however, the grain of the wood demonstrates the reality. There is a good

---

④ 拓扑学：研究在变形（弯曲、拉伸、挤压，不含破坏）作用下不发生变化的几何物体的属性。
Topology: In mathematics, the study of the properties of a geometric object that remains unchanged by deformations such as bending, stretching, or squeezing but not breaking.

⑤ 在彼得堡大教堂中，采用了类似的从实心柱向楔形块的转化策略。这些楔形块在几何上边界未对齐，但在下边界（可视面）精确对齐。
TA similar strategy of the solid column transitioning into voussoirs above was employed in Peterborough Cathedral. These voussouirs also misalign on the upper bound geometry, while aligning precisely on the lower bound (visible surface).

⑥ 侧刃铣削是沿构件表面（如锥形脊侧壁）行进时使用立铣刀进行侧向切削的技术。
Swarf machining is a technique that allows side cutting with an endmill while proceeding along the surface of a part, such as the sidewalls of a tapered rib.

⑦ 菲利贝尔·德洛姆（16世纪）和帕拉迪奥一样，都是石匠的儿子。他在建筑领域为人们所熟知并非因为对形式或技术的深入了解，而是因为其建造师，或石匠的身份。在1567年出版的《第一本建筑》中，德洛姆介绍了艺术几何特征的方法和定义。该方法指导我们如何从图纸到建筑，及从建筑到图纸。因为这本书，德洛姆也被视为第一位专业建筑师。尽管，当代的建筑再现方式和德洛姆的作为建造方法模板的画法几何之间存在很大的区别，但他发明的技术使得当时的设计师与建造师之间能够进行指导和沟通。因此，德洛姆可以说是数字建造的鼻祖。
Philibert de L'Orme (Sixteenth Century) was, like Palladio, the son of a mason. He emerged into architecture, not through a series of rigorous understandings of form or technique, rather from the builder—or mason. In his printed work of 1567 Le premier tome de l'architecture, Philibert de L'Orme introduced the method and definition of art du trait géométrique. This method developed as a way to reciprocally draw what can be built and vice-versa. Because of this emergence, De L'Orme can also be credited as the first professional architect as his technique served to instruct and communicate between the designer and the builder, though an important distinction should be drawn between the representation of architecture we now generate and De L'Ormes descriptive geometry that served as method template to construct. In a way, De L'Orme can be considered the predecessor to digital fabrication.

结论

LeFevre拱证明了将围绕体量的当代建造方法和古代知识相结合的可能性。它成功地采用了物理模拟，借助体量计算和体量制作流程的协同，确保了结构的稳定性。在此案例中我们采用了聚合式波罗海桦木胶合板作为一种模拟，但我们仍旧能够看到其他体积材料，如蒸压加气混凝土、石膏或石材运用在这一领域的潜力。

拱的组装
Assembly of the vault

reason for this false-reality. A column does not perform in the same manner as a vault. The thrust-vectors inside the column are vertical, not progressively horizontal. To that end, a column does not resist horizontal thrust. It is resists buckling. The solidity of the column is paramount.

The discrepancy in transitioning from solid column to discretized vault is resolved via rhetoric. The rhetoric of individual units continues down the column as if the single and solid column was in fantasy an impossible continuation of the units to the ground. This rhetoric is not a simple continuation of the conical-boolean geometry that composes the vault. It is a new, yet similar approach. It refers to the conical-boolean, without repeating it. This shift in geometry allows the system not only to calibrate volume (as applied in the vault), but also to perform another transition from fragmented to smooth. As the units make their way down the column, they do get smaller, but the dimples slowly make their way to the surface producing the illusion of continuity, only to push through that continuity as the very base. This punctuation to the statement suggests that the weight of the vault above is so great that the column is forced to bulge outward.

Fabrication

The vault is produced with Baltic Birch plywood. The plywood is sourced in 19.05mm thick sheets awaiting the 'thickening'. Perhaps this speaks to the state of the industry that volumetric material is difficult to procure. Each custom unit is digitally dissected and sliced into these thicknesses, cut from the sheets, and then physically re-constituted into a rough volumetric form of their final geometry. These roughs are indexed onto a full sheet and glued, vacuum pressed, and re-placed onto the CNC router. This process is materially more efficient than carving these units from one solid block of material, though it is more laborious.

This project is produced on a 5-axis Onsrud router. The swarf [6] tool-paths utilized are dedicated to removing the most material with the least effort. Instead of requiring the end of the bit to do the work, this path uses the edge of the bit to remove much more material. Because this method traces the geometry with a line as opposed to point via Philibert De L'Orme's technique stereotomy [7], it requires the units be constituted of ruled surfaces. This constraint informed the conical-boolean geometry in the vaulted portion of the project, though relaxed in the columns where a more typical surface milling operation produces the rhetorical bulges. This shift in tooling operation also speaks to the understanding of difference between column and vault.

Analysis

This project was fabricated with an assumed zero-fill approach. As part of the requirement that the vault must be dismantled, there is no mortar. Discrepancies, errors, and gaps were impossible to resolve because or this zero-tolerance approach. In order to ensure completion on site in difficult locations, a manual band saw handled the work of removing collision material on the backside of the problematic units. This site carving did not affect the front edge of the units, but it did produce a gap where the voussoir surfaces were not coincidental. This happy accident aligns precisely with the Inca wedge [13] process where masons would fill from the backside of a wall with mortar into a voided wedge between stones, while the front and architectural face appeared to be mortar-less. There is room for further exploration to capitalize on the potential of the Inca wedge method.

Conclusion

La Voûte de LeFevre demonstrates the potential of informing contemporary fabrication methodologies with past knowledge surrounding volume. It successfully employs physics simulation to ensure stability through volumetric calculations that serve in reciprocity with volumetric making processes. While aggregate baltic birch plywood serves as an analog, we see potential in other volumetric materials such as autoclave aerated concrete, plaster, or stone.

参考文献 / References：

[1] Block, Philippe, Thierry Ciblac, and John Ochsendorf. "Real-time limit analysis of vaulted masonry buildings". Computers and Structures 84, 29-30 (2006): 1841-1852.

[2] Fry, Ben and Casey Reas. "Processing web site" , 2011, Accessed 01 November 2011. <http://www.processing.org/>

[3] Kilian, Axel and John Ochsendorf. 'Particle-Spring Systems for Structural Form Finding' , Journal of the International Association for Shell and Spatial Structures. 46, 2 (2005): 78.

[4] Ibid., 77-84.

[5] Heyman, Jacques. "The Stone Skeleton". International Journal of Solids and Structures 2 (1966): 249-279.

[6] Block, Philippe and John Ochsendorf. "Lower-bound analysis of unreinforced masonry vaults' . In Enrico Fodde, ed. Proceedings of the VI International Conference on Structural Analysis of Historic Construction, CRC Press (2008): 593-600.

[7] Otto, Frei and Bodo Rasch. Finding Form: Towards an Architecture of the Minimal. Berlin: Edition Axel Menges, 1995.

[8] Kilian, Axel and John Ochsendorf. 'Particle-Spring Systems for Structural Form Finding' , Journal of the International Association for Shell and Spatial Structures. 46, 2 (2005)

[9] Clifford, Brandon. 'Thick Funicular: Particle-Spring Systems for Variable-Depth Form-Responding Compression-Only Structures' , In Xavier Costa and Martha Thorne, ed. Change Architecture Education Practices: 2012 ACSA International Conference. New York: ACSA Press (2012): 475-481.

[10] Clifford, Brandon. 'Thicker Funicular: Particle-Spring Systems for Variable-Depth Form-Responding Compression-Only Structures' , In Paulo Cruz, ed. Structures and Architecture: Concepts, Applications and Challenges. London: CRC Press (2013): 205-206.

[11] Block, Philippe, Thierry Ciblac, and John Ochsendorf. "Real-time limit analysis of vaulted masonry buildings". Computers and Structures 84 (2006): 29-30.

[12] Ibid

[13] Clifford, Brandon. Volume: Bringing Surface into Question. SOM Prize Report, SOM Foundation, 2012, pp. 286-289.

阴影线
2014年欧洲太阳能十项全能竞赛项目的原型建造

指导老师：鲍勃·谢尔，伊曼努尔·沃科鲁兹，凯特·戴维斯
学生：盖里·艾德瓦兹，格里格·格里戈罗夫，尼根·阿弥利达哈，卢克·鲍威尔，约书亚·布罗姆，尼古拉斯·德布鲁恩，马修·胡德斯皮斯，格林·伍德里奇，张亮

Shadow Lines
Shelter Prototype of 2014 Solar Decathlon Competition (Versailles)

Tutors: Bob Sheil, Emmanuel Vercruysse, and Kate Davies.
Students: Gary Edwards, Grigor Grigorov, Neguin Amiridahaj, Luke Bowler, Joshua Broomer, Nicholas Debruyne, Matthew Hudspith, Glenn Wooldridge, Liang Zhang

Shadow Lines
An Experiment in Simulated and Speculative Manufacturing
阴影线
一种模拟和推测性建造的试验

Bob Sheil, Thomas Pearce, Grigor Grigorov / The Bartlett School of Architecture, UCL
鲍勃・谢尔，托马斯・皮尔斯，格里格・格里戈罗夫 / 伦敦大学学院Bartlett建筑学校

这仅是全球许多学生针对自动建造开展的众多设计实验之一，是教育界在处理研究和工业的关系上发生的根本性变革的一个表现。在此情况下，它不是一个强调任何重大意义的物体，而是作者为了传达本质而必须掌握的专业知识。它不是一个项目，不是一座大楼，不是一个结构，而是一个探索性的原型。作为一个物体，它被视作无关纸笔的试验图纸，好比手工艺品，从构思到执行的整个过程中挑战了学生作为一个设计师具有的潜力和预期。因此，机器人的价值不在于其敏捷度、适应性或稳健性，而是对即将复兴的传统设计行业带来的深远影响。

This demonstration is only one of many experiments in design through automated manufacturing that many students across the world are researching today. It is part of a radical transformation of education's relationship with research and industry. In this instance, it is not the object that claims any great significance, but the spectrum of expertise its authors had to straddle and command in order to deliver what it is: not a project, not a building, not a structure, but a speculative prototype. As an object it is to be read as an experimental drawing, not only without paper, but without pens, and as such an artifact that challenges the students' expectations on where they sit as a designer on the pathway from conception to execution. The value of robotics therefore lies not in their dexterity, adaptability, or robustness, but in their status as a profound provocation towards a deeply conventional design industry that faces a new renaissance.

2003年伦敦大学学院Bartlett建筑学校成立了MArch Unit 23工作室，作为理学学科的一次改革，MArch Unit 23始于1997年由鲍勃・谢尔和尼克・加里考特组织的一次学校的设计工作营。两人于1995年共同创立了Sixteen*公司（制造商），在学院中率先担任这个工作营的指导教师。如今，在鲍勃·谢尔，伊曼努尔・沃科鲁兹和凯特・戴维斯的领导下，这个工作室积极试验并探索建筑生成方式的新手法。《建筑设计中的计算机辅助建造》（由建筑出版社在2000年出版）是最早发表的有关数字建造的著作之一。这本书的作者尼克・加里考特在2003年辞去了学术职务，与克里斯蒂娜・厄勒托一起成立了致力于德国钢铁行业数字化建造流程的Ehlert-Stahlbau GmbH公

MArch Unit 23 was formed in 2003 at The Bartlett School of Architecture UCL as a remould of a BSc undergraduate unit that began in the School's workshops led by Bob Sheil and Nick Callicott in 1997. Both Sheil and Callicott had established the practice 'sixteen*(makers)' in 1995 and were the first design tutors in the school to teach out of its workshop. To the present day, now led by Bob Sheil, Emmanuel Vercruysse, and Kate Davies, the unit has actively pursued an experimental approach towards how architecture is made and by whom. Author of one of the first publications on digital fabrication, *Computer-Aided Manufacturing in Architecture* (Architectural Press 2000), Callicott

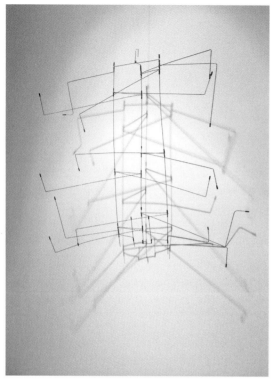

"无纸的画作", 2015年在巴黎的拉丁美洲之家中展出的景象（格特鲁德·高德斯米德 1912~1994）作品
"Drawing without Paper" Gego (Gertrud Goldschmidt 1912~1994) at Maison de l'Amerique Latine Paris 2015.

线是由一系列连续的点组成的
A line is a dot that went for a walk

——保罗·克利 Paul Klee

司。鲍勃·谢尔随后逐步改革了Bartlett学校的设计工作营，并于2013年投资200万英镑创办了Bartlett建造与设计交流中心（B-made）。与之毗邻的MArch Unit 23工作室当前拥有11套机械臂和全英高校中最齐全的建成环境领域内的设计和建造工具。本文主要介绍MArch U23工作室最近在B-made中承担的主要项目。

随着增材制造迅速占据消费品市场，大众和建筑出版物一致认为增材制造具备前所未有的创造潜能。但是需要认识到，任何新兴技术在变革生产机制和层次结构的同时，也在自动化生产的过程中产生了新的定式行为和思考方式，它的变革性质也因大打折扣。增材制造为当代生产和制造概念带来的其中一种思维方式就是越来越强调"外皮"或者"外壳"是室内和室外空间之间的明确分隔。

left academia in 2003 with Kristina Ehlert to establish 'Ehlert-Stahlbau GmbH' a business in the German steel industry devoted to digital processes. Sheil subsequently led a gradual reformation of the Bartlett's workshops with £2m of investment launched the Bartlett Manufacturing and Design Exchange (B-made) in 2013. MArch Unit 23 locates itself in close proximity to B-MADE, which is currently equipped with 11 robotic arms and the largest array of design and fabrication tooling in any UK Higher Education Institution in the Built Environment sector. This paper is an overview of a recent project undertaken by MArch U23 at B-made.

As additive manufacturing rapidly takes charge of consumer markets it is hailed in popular opinion and architectural press as revolutionary in its unforeseen proliferation of creative potential. It is important to recognize, however, that every emerging technology, while undeniably transforming mechanisms and hierarchies of production always simultaneously generates, in the process of its automation, new stereotypical, and hence rather less "revolutionary" forms of behavior and modes of thinking. One of the modes of thinking smuggled into contemporary concepts of production and fabrication by additive manufacturing seems to be a growing fixation on the notion of the "skin" or the "shell" as a discrete and unequivocal separation between inside and outside.

The automated black box of the 3D printer and its software require of an object that it is "watertight", that it has no flaws or "leeks" in the geometrical definition of inside and outside. With this enforced fixation on the clear cut shell, a certain sense of indeterminacy and intuition gets lost within the process of making – even if the objects printed might often mimic the very notion of generative indeterminacy. Simultaneously, it engenders a more ambiguous understanding of the topological-architectural threshold, closing off the thick and negotiable edge that defines a permeable architectural skin.

Based on a competition brief to design a temporary shelter for the 2014 Solar Decathlon in Versailles, the project "Shadow Lines", designed and prototyped by students of the Bartlett's MArch Unit 23 and lead by Bob Sheil, Emmanuel Vercruysse, and Kate Davies, tries to re-appropriate this indeterminacy and ambiguity both through its approach to the process of additive manufacturing and by means of the architectural output it proposes. Using small-scale 3D printing and robotic plastic depositioning, it establishes a liminal ambiguity by re-imagining additive manufacturing as an accumulative and iterative drawing process. The proposed design, rather than acting as a closed shell, is drawn as a semi-permeable network of 3 dimensional hatch lines that mediates between the sun and the user, modulating the sun rather than blocking it, bending and diffracting the solar rays rather than reflecting them.

3D自动打印机的黑盒子和软件要求对象必须"防水"，几何体内部和外部都不能有任何瑕疵或者"裂缝"。这种强制制造出的外壳轮廓分明，导致在制作过程中的不确定性和直觉消失，即使打印对象常常会模拟"生产中的不确定性"。同时，这种打印方式危及了拓扑和建筑学之间的模糊界限，取消了边缘的厚度和可协调性，从而定义了一种不可渗透的建筑表皮。

在法国凡尔赛市举行的2014年欧洲太阳能十项全能竞赛的要求设计一个临时庇护所。由Bartlett建筑学校MArch Unit 23 的学生进行原型设计，鲍勃·谢尔，伊曼努尔·沃科鲁兹和凯特·戴维斯主导设计了"阴影线"项目。项目试图通过结合使用增量制造和建筑输出两种方式重新寻找不确定性和模糊性。采用小型3D打印和机器人塑料沉积法来确定模糊阈值，将增量制造流程重新设想为不断绘图的过程。设计方案并非封闭的壳体，而是绘制为半透水的三维剖面线系统，关联了用户和阳光之间的关系。调节而非遮挡阳光；折射和衍射而非反射阳光。

我们在桌面3D打印机构建的美观脚手架造型的启发下，设计了这些具有逻辑性和美观特征的重叠线，进而为打印对象不透明的外壳提供了支持。尽管在正常情况下，市面上可购买的桌面3D打印机套装软件可自动生成这些脚手架造型，但是，本项目试图对该生成过程进行设计控制。为此，我们有必要探索内在算法过程，以将输入的几何图形转换成切片，并最终转换成G代码。这十分适合于3D打印"外皮"，如黑匣子用户界面。

我们采用自建开源数字桌面3D打印机，通过直观的手动"涂写"笔流程进行小规模测试，并开启塑料沉积流程及对其作为"画图"媒介的潜能的初步探索。我们已经对不同的形式生成和打印方法进行了探索，包括传统3D建模及由颜色图引导的基于脚本的刀具轨迹生成；采用了计算模型以直接生成G代码，这使得可通过设计流程控制每个输入机器的移动命令，突破了严谨的切片机及刀具轨迹生成软件的局限。在塑料挤压机系统的适配下对机器可读代码生成的控制可确保很容易利用桌面3D打印机将其转换成6轴工业机器人，并使其具有可扩展性。

The logic and aesthetic of these overlapping line patterns originates from a fascination with the accidental beauty of the scaffolding patterns that are created by desktop 3d printers to support the opaque shell of the object intended for print. While these scaffolding patterns are normally automatically generated by commercially available desktop 3D printer's software packages, the project seeks to gain design control over this generation. For this, it is necessary to explore the underlying algorithmic processes that translate the input geometry to slices and eventually to G-Code – thereby also quite appropriately looking beneath the "skin" of the 3D printing process itself – in other words, its blackboxed user interface.

Initial explorations of the plastic depositioning process and its potential as a "drawing" medium began with small-scale testing using self-built open source digital desktop 3D printers and the intuitive process of the manual "scribbler pen". Different methods of form generation and printing were explored ranging from conventional 3D modelling to script-based tool-path generation guided by colour maps. Computational models are used to generate the G-code directly. This allows the design process to control every move command passed to the machine, removing the limitations of rigid slicer and tool path generation software. The control over the generation of the machine-readable code guaranteed an easy translation and scalability of its application from the desktop 3D printer to an industrial 6-axis robot fitted with a plastic extruder.

Initially using lines generated between random points, the generation and fitness algorithms became more refined to include the shape of the bounding box, openings for supports and using gradient maps for the computation of the density of the generated line web. Using open-source as well as custom-built software plug-ins, a high level of control over the generation of the scaffold is created so that eventually the figure ground relation between

 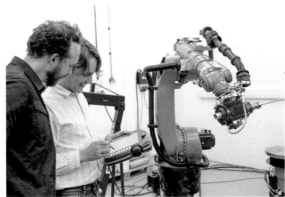

"阴影线"项目的建造是在Bartlett建造与设计交流中心（B-made）的支持下完成的
"Shadow Line" project was supported by Bartlett Manufacturing and Design Exchange (B-made)

   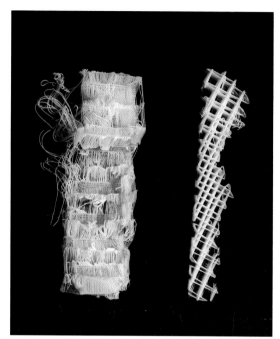

"阴影线"项目的原型及网格纹理推敲
Networked Patterns Analysis, Shadow Line, 2014

首先，我们采用了任意点之间生成的线条，使得生成及适应值算法更加优化，并可塑造边界框造型及支撑开口。其次，采用渐变映射计算生成的线网密度。采用开源及定制软件插件可高度控制脚手架的生成，并最终颠倒脚手架和外壳之间的图底关系。其中，外壳仅用作脚手架的计算容器，并非真正打印，而其设有"外皮"的填充物成为真正的设计对象。

庇护所原型的设计及其网络模式的生成，反复响应一系列基地、业态及生产的具体指标。在欧洲太阳能十项全能竞赛的大背景下，太阳自然而然成为了最重要的限制条件。参数化日晒量及遮阳分析软件生成渐变映射，对应节日具体时间段内的太阳轨迹，最终确定生成线条图形的密度及形状。同样地，凡尔赛公园内的基地为设计避难所外皮的半渗透肌理提供了灵感，响应了其诸多迷宫中叶子的多孔性特点。此外，生产及交通的限制条件约束且规定了面板大小、挤压角度等。通过嵌入部件的整合（按照其在完成结构上的位置整合到挤塑板），能够启用基地面板的接缝。

单独面板的大小旨在便于搬运，以方便包装并运至基地。采用空心截面的铝管提供水平结构，并将面板重叠穿过铝管，塑料面板自身的结构强度构成垂直结构。采用逆电流器作为分隔部件，以增加覆盖深度，而各编码开洞的变化为施工过程中的一些展示提供了机会。每块面板约$1m^2$。渐变映射说明了塑料开放网络结构的编码密度及印刷密度。材料体积比得到了优化。展馆布局十分灵活，且可根据预算、尺寸要求及工期进行调整。

scaffold and shell is inverted, in that the latter is merely used as a computational container of the scaffold but isn't actually printed, its "skinned" infill becoming the actual object of design.

The shelter prototype's design and the generation of its networked patterns iteratively react to a series of site, program and production specific parameters. Within the context of the Solar Decathlon, the sun naturally provided the most important constraint. Parametric insolation and shading analysis software created gradient maps, corresponding to the sun-path of the specific time period of the festival, which then defined the density and shape of the generated line patterns. Also, the site within the park of Versailles was inspirational for the creation of the semi-permeable texture of the shelter's skin, echoing the porosity of the foliage of its many mazes. Furthermore, production and transportation constraints limited and dictated panel sizes, extrusion angles etc. The joining of the panels on site was to be enabled by the integration of embedded components that were integrated in the plastic extruded panels according to their position in the finished structure.

Individual panels are sized for easy handling and can be easily packed for shipping to the site. Hollow section aluminium tubes provide the horizontal structure, and the panels are threaded onto them in overlapping layers. The structural strength of the plastic panel itself provides the vertical structure. Spacers are used as

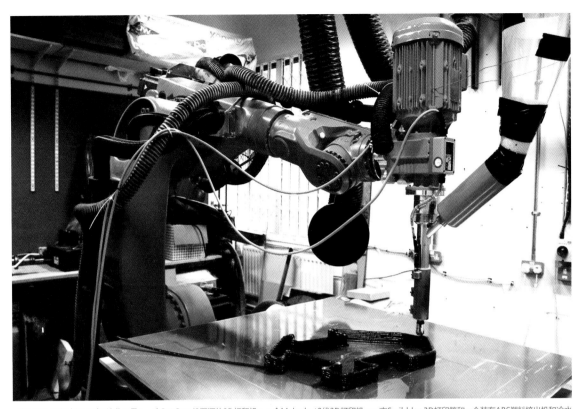

"阴影线"项目中运用到了这些工具：一个RepRap i3开源的3D打印机；一个Makerbot2代3D打印机；一支Scribbler 3D打印笔和一个装有ABS塑料挤出机和冷水机组热板的KUKA KR60的机器人
The following tools were deployed in the project: A RepRap i3 open source 3D printer; Makerbot Replicator 2X 3D printer using modified "slicer" machine code; Scribbler 3D printing pen test drawings; KUKA KR60 with ABS plastic extruder with hot plate and chiller unit.

塔楼作为日晷，阴影端部于特定时间落在长凳卫星上。塑料元件与嵌入部件共同使用，发挥路标及日晷的作用。潜在座椅元件将用作计时器，如此一来，用户将发挥影子的作用，也同时实现了嵌入元件的共同使用需求。从而我们得出了想要的结论，而脚手架也在各不同元件之间建立了链接。

separating members to increase the depth of the overlay, and the variation of encoded holes provide an opportunity for some play in the construction. Each panel is approximately 1 meter square. The gradient map represents the density in which the plastic open web structure is coded and printed. The volume to material ratio is optimized. The arrangement of the pavilion is flexible, and can be adjusted for budget, size requirement, and time.

The tower acts as a sundial with the tip of its shadows falling on sitting bench satellites at particular times. Plastic element joined with embedded components to be used as a landmark and sundial. Potential seating elements would be used as time markers. (The user is invited in such a way as to contribute to the shadows as well.) Joints using embedded components were made. Outlines are drawn, the scaffolding emerges between the elements.

最终成果
Final Result

Clay Robotics
机器人粘土打印

Kwang Guan Lee, Sun Jiashuang / The Bartlett School of Architecture, UCL
康源，孙佳爽 / 伦敦大学学院Bartlett建筑学校

机器人粘土打印技术通过使用6轴工业机械手臂，结合现有混凝土铸造技术与叠层制造流程，提出了一个新的制造系统。这个首次提出的方法使用粘土作为可持续、环保和低成本的成型模具材料。这种材料可用性强，而且过滤和再利用的成本很低。并且粘土是为数不多的几个建筑材料之中，可以很容易地通过叠层制造①操纵。总而言之，机器人粘土打印结合了流体挤压（叠层制造的一种形式），以及机器人技术的精度和自动化。使用粘土作为模具材料，进而模具可以被打印以及塑形成复杂的几何形体。本研究的目的是开发一个可行的混凝土筑造技术，以1：1的比例进行，比传统的建造更加可持续且更有效。

材料特性与研究

粘土作为最古老的建筑材料之一，在5000多年前就被用于各种手工艺品和砖的制造。当与水混合时，粘土混合物可以被用来塑型，当水分蒸发后，其又会渐渐变硬。这一特点使得它很适合被

Clay robotics is a proposed fabrication system that hybridizes existing concrete casting techniques with additive manufacturing processes through the use of six-axis industrial robotic arms. It is a speculative approach on using clay as a sustainable, environmentally-friendly and low-cost casting mold material. The material is widely accessible, and the cost of filtering and treating to be used for construction is low. In addition, clay is one of the few building materials that can be easily manipulated through additive manufacturing ①. In sum, Clay Robotics combines paste extrusion, a form of additive manufacturing, with the precision and automation of robotics technology. Using clay as a mold material, a formwork can be printed and shaped to allow for free-form casting of complex geometries. The objective of this research is to develop a viable concrete casting technique at 1:1 scale construction that is potentially more sustainable, and efficient than traditional construction.

① 叠层制造是现代产品开发的一个组成部分。这项技术的潜力在于材料能以连续的方式沉积，类似于有机增长的概念。
Additive manufacturing is now an integral part of modern product development. The potential of this technology is in the way material is deposited in successive layers, similar to the notion of organic growth.

用来制作复杂的几何形体，并在其变硬后被用来作为模具使用。从最初筑型测试开始，我们制作了一系列的物理模型来模拟内腔模型的形成过程。这就类似于白蚁丘的内腔模型一样，只不过我们将土壤替换成粘土。最终证明错综复杂的模具可以被用来浇筑混凝土。与传统混凝土浇筑技术相比，临时的模具通常是由木材、铝或塑料组成的，然而使用粘土的一个明显的优势就是，它不局限于任何形式，我们能够使用粘土塑造任何有机的、不规律的几何形体。很显然，使用粘土作为模具材料，其粘度是决定模具强度、表面光洁度、高度和形状的主要因素。因此，我们做了几个测试来找出塑造模具和浇注混凝土的合适粘度，确定最佳水和粘土的混合比。同时，在挤压粘土的过程中，我们也发现不同的移动速度以及喷头打印角度和高度也直接影响了打印的图案，所以我们也通过控制变量的方式，将打印中出现的自然叠落图案做成了目录，方便之后的参数读取。此外，在倒入混凝土之后，粘土模具可以很容易通过剥离，通过在水软化，或溶解在温和的酸性液体的方式将其去除；甚至是模型里面最难碰触的地方，模具也可以被清理干净。由于粘土模具在任何方向都能被溶解和去除，这解决了传统建筑中混凝土去膜的问题。同时，一旦粘土被除去，用过的模具可以通过过滤和水混合被重复利用。相同的粘土可以用于塑模，也可以在浇筑之后重新使用，从而形成了材料使用的循环系统。

工具开发及挤压试验

流体打印头是3D打印工具的一种类型，其可针对例如像粘土这样的膏状材料，进行自动的分层堆叠打印。为了实现能按照1∶1的比例打印模具，并可以通过机械手臂进行精密控制的目的，我们在容积式系统的基础上，针对高粘度的粘土开发了末端打印头。标准的容积式挤出头有两个主要组件：泵和挤出机。最终的样品是由一个小尺寸的螺杆转子和软体橡胶定子组成。这可以对喷嘴起到密封作用，通过吸力可以保持材料不流动，从而成为材料挤压机的开关。这将更加精确地掌握材料挤出机的启动和停止，挤出速度和所沉积的材料数量。

我们尝试了两种3D打印工具的设置方法。第一种是以底座为中心的设置方法，其拥有固定的底盘，和一个可以遵循指定的路径移动的挤出喷口。它的主要优点是只要泵可以将材料推进挤压管的长度，模型便可以随意调整比例；而这个方法的缺点就是粘土本身强大的后推力(例如摩擦力，材料自身的重量)，均可以将材料反向推回挤压机里。以底座为中心的设置方法，我们用含水量更多的粘土进行了几个挤压测试，从而研究粘土的堆叠效果。然而，由于粘土自身的流动性和重量，打印的粘土越高，其自身的变形就越明显。为了获得一个更大尺寸上的样品，通过把模型分成单独的组件，我们制作了一个直径50cm，高70cm的石膏模型。同时，我们也制作了一个木制框架支撑在粘土模型周围，并且在粘土与木框之间填充沙子，以确保在注入石膏时不会有泄漏的地方。另一种设置方法是以喷嘴为中心，拥有固定的喷口和可以遵循指定路径移动的底盘。其主要优点是材料可以以更短的距离被挤压出来，这也就意味着可以

Material Characteristics and Research

As one of the oldest building material, clay was used to make various artifacts and bricks over 5000 years ago. Clay can be shaped when mixed with water, and hardened when moisture is evaporated from it. This makes it suitable for manipulating into complex geometries when it is less viscous, and then stiffened to be used as a mold. From initial casting tests, a series of physical models were made to simulate the endocasting process. Similar to the termite mound endocasting, soil is substituted with an earthy material like clay, an intricate mold can be created for casting. Compared with traditional concreting, where a temporary formwork is usually made of timber, aluminium or plastic, one clear advantage of using clay is that it is not as restricted to any shape, enabling it to take on any organic, non-standard geometry. Obviously, using clay as the casting mold material, its viscosity is main determinant of the strength, surface finish, height and shape of building mold. Therefore, we did several tests to find out the appropriate viscosity for shaping a mold and casting concrete, determining optimal water to clay mix ratio. At the same time, in the process of extrusion clay, we also found that the printing pattern of clay mold is affected by the different moving speed, angle of nozzle and the height between the nozzle and base. So we used the way of control variables to summarize the printing pattern into a catalogue, which is convenient to read parameter later. Besides that, after the cast is poured, the clay formwork could easily be removable either by peeling, by softening in water, or by dissolving in a mildly acidic liquid. The formwork is washed away even in the most inaccessible spaces. This alleviates the issues of formwork removal in conventional construction, as the formwork could be dissolved and removed in any direction. Once removed, the clay can be reused through filtering and mixing into water. The same clay can be used to shape the mold and reused after the cast, forming a closed-loop system.

Tool Development and Extrusion Tests

A paste extruder is a type of 3D printing tool that can be used for automated layer deposition of paste-like material, such as clay. To print the formwork in clay at a 1:1 scale with robotic-controlled precision, we developed a prototype end-effector extruder for material of high viscosity based on a positive displacement system. The typical positive displacement extruder is set up with two major components: the pump and the extruder. The final prototype is built with a small-scale progressive cavity rotor with a soft-body rubber stator. This would create a seal at the nozzle, holding the material in by suction, which would act as a valve for the material extrusion. This will allow precise starting and stopping of the extrusion, as well as the speed and amount of material that is being deposited.

打印参数分析
Printing Parametric Analysis

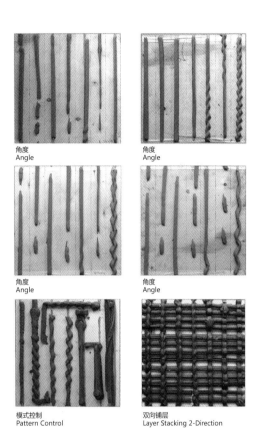

角度 Angle | 角度 Angle
角度 Angle | 角度 Angle
模式控制 Pattern Control | 双向铺层 Layer Stacking 2-Direction

使用更高粘度的粘土；而缺点是这种设置方法不容易轻易改变比例大小。最终我们确定，以喷嘴为中心的设置方法同时结合一个螺杆泵，是打印大型粘土几何形体的最理想组合。此外，与先前的设置方法相比，粘土粘度从低到高的提升对于流体挤压机来说是巨大的进步。作为测试以喷嘴为中心的设置方法能够建造的最大模型尺寸，我们打印了一个直径为300mm高为500mm的圆柱状体。此模具厚壁为2cm，并留有一天时间自然风干，以便浇筑。以此，粘土模型证明了其强度本身足够坚硬来抵抗浇筑混凝土带来的静压力，并由于粘土的层层堆叠，我们得到了混凝土带纹理的表面效果。

柱式原型

在最后一个学期，我们将测试高度、悬垂和跨接的柱子均加入钢丝进行浇铸。铸造后，通过上下堆叠的方式，搭成了一个2m高的柱子。因为模具没有完全去除水分，所以柱子的每个部分分别浇铸以防止粘土模具塌陷。柱子的形式从底部一个直径300mm的圆逐渐向上扭转，分成6个独立的分支，并在柱子的顶端再次合并在一起。为了推进粘土挤压系统的极限，我们建造了一个高1.4m，带有7个分支桥接的复杂扭曲的混凝土柱子。这种几何形体就类似于白蚁穴内部的各个空腔一样，每个分支都从中心结构中分离而后连接起来。接着，在让粘土模

Two different setups of 3D printing tool have been tried. The first setup is called base-centric setup, which has a stationary base, and a moving extrusion nozzle that follows a designated toolpath. The main advantage is that it is completely scaleable, given that the pump can push the material through the length of the extrusion pipe. The disadvantages are the strong forces that push the material back against the extrusion (i.e. friction, weight of material). With the base-centric setup and a wet mix of clay, several extrusion tests were made on studying how the layers of clay will stack. However, due to the fluidity and weight of the clay, there were large deformations at the base of the object as it was printed higher. To achieve a prototype at a larger scale, by splitting the model into separate components, we built a plaster cast model with a diameter of 50cm and a height of 70cm. At the same time, a wooden frame was constructed around the clay mold, and infilled with sand to ensure that there were no leakages when pouring the plaster. The other setup is called nozzle-centric setup which has a stationary extrusion nozzle, and a moving base that follows a designated toolpath. The major advantage is that the material is extruded at a much shorter distance, which means the higher viscosity clay can be used. The disadvantage is that this setup not easily scaleable. Ultimately, the base-centric setup with a progressive cavity pump for viscous material is ideal for

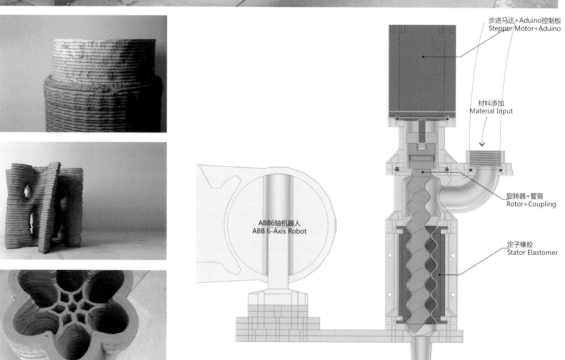

机器人建造过程及打印工具头
Robotic Fabrication Process & Printing Tool

具风干了一天之后，我们将每个模具组件都组装在一起，在将200kg的混凝土一次浇灌在模具里。为了防止每个模具部分之间搭接的位置会有漏缝，我们用石膏去修补模具之间的衔接处。这个模型很明显的展示了使用粘土作为模具材料的优势，同时也展示了用流体挤压机来建造这样复杂几何形体的高效。

应用

回想整个建筑历史，建筑形式的表达一直受限于传统的建筑方法，哪怕是轻微的不规则的形式也可以造成施工所需时间和成本的大大浪费。然而机器人粘土打印——这种沉积建造的方法，可以用制作一个很普通的几何形体所需等量的时间和成本，来构建一个曲线的不规则几何形体。在一些地区，建筑材料例如水泥、木材和石头并不易于得到，而粘土却是当地特有的材料，它可以直接用于机器人粘土打印，去建造低成本、可重复使用的构建组件。小规模难民收容所或者自然灾害的受害者临时住处都可以使用这样的组件构建以解决住房紧缺的问题。就像白蚁穴可以根据通风或光照去改变蚁穴的几何形态一样，机器人粘土打印也可以实现通过复杂的模具结构来准确的应对建筑周围微环境。

机器人粘土打印为建筑自由形体的建造提供了可能
Freeform Construction

项目团队 Project Team

指导教师：菲利普•莫雷尔，蒂博•施瓦茨，康源
Instructors: Philippe Morel, Thibault Schwartz, Kwang Guan Lee
学生：孙佳爽，凯尔文•何，王思涵
Students: Sun Jiashuang, Kelvin Ho, Wang Sihan

large-scale clay printing. Also, compared with previous extrusions, the move from a low viscosity to a high viscosity mix was a huge step to better paste extrusion. As a test for the maximum build size with the nozzle-centric setup, we extruded a straight round cylinder with a 300mm diameter to half a meter in height. The mold was built with a 2cm thick wall, and allowed to dry for a day before casting. The clay mold proves to be rigid enough to hold the hydro-static pressure of the cast, and leaves a subtle grooved surface finish left from the layers of clay.

Column Prototype

In the final term, the column tests for height, overhang and bridging are casted with material wire reinforcement. After casting, they are connected by stacking one on top of the other to form a two meter high column. The column sections are casted separately to prevent the clay mold from collapsing, as they were not left to dry completely. The writhing geometry splits from a 300 mm diameter wide base to six separate branches, and merging back at the top of the column. To push the limits of the clay extrusion system, we built a complex twisting column with 1.4 meters high and seven bridging branches, which would otherwise be difficult to construct using traditional formwork. Similar to the internal cavities of a termite mound, the branches arch and connect to a central structure. Later, after letting the clay dry for a day, the formwork was assembled to full height and 200kg of concrete mix was poured in one go after. To prevent leakages between each mold section, plaster was used to patch the connections. This model clearly demonstrates the advantages of using clay as a formwork material for casting, and the efficiency to build such complex geometry with paste extrusion.

Application

Throughout the history of architecture, the expression of building form has always been limited by traditional building methods, in which even slightly irregular forms can significantly change the time and cost needed for construction. However this layer deposition building method would take an equal amount of time and cost to build curved irregular geometry vs a straight regular geometry. In regions where alternative building materials, such as cement, timber or stone, are not available, and where clay is vernacular, it can be deposited directly in layers to form low-cost, reusable building components. Small-scale shelters for refugees or victims of natural disaster can be built using these components to address the issues of emergency housing in need. Complex formwork allows for the realization of structures that react to its environment at precise micro-scale, much like how termites change the geometry of their habitat according to wind ventilation or sunlight.

设计团队

斯图加特大学
计算设计学院：门格斯教授，托比亚斯·斯科文，奥利弗·大卫·克里格
结构设计学院：尼佩尔斯教授，李建民
工程测量学院：沃尔克·施维格教授，阿妮塔·施密特

Müllerblaustein木结构公司
瑞赫·穆勒，本杰明·埃塞尔

KUKA机器人公司
阿洛斯·布赫夕塔布，弗兰克·齐默尔曼

Design Team

University of Stuttgart
ICD Institute for Computational Design: Prof. A. Menges (PI), Tobias Schwinn, Oliver David Krieg
ITKE Institute of Building Structures and Structural Design: Prof. J. Knippers, Jian-Min Li
IIGS Institute of Engineering Geodesy: Prof. Volker Schwieger, Annette Schmitt

Müllerblaustein Holzbau GmbH
Reinhold Müller, Benjamin Eisele

KUKA Roboter GmbH
Alois Buchstab, Frank Zimmermann

Landesgartenschau展馆夜景，室内的照明强调了一分为二的主要建筑空间。
Night view of of the "Landesgartenschau Exhibition Hall". The interior lighting accentuates the building's spatial organisation into two main spaces.

Landesgartenschau Exhibition Hall
Robotically Fabricated Lightweight Timber Shell
Landesgartenschau 展馆
机器人建造的轻型木壳结构

Tobias Schwinn, Oliver Krieg, Achim Menges / ICD Institute for Computational Design, University of Stuttgart
托比斯·斯科文，奥利弗·克里格，阿希姆·门格斯 / 德国斯图加特大学计算设计学院

Landesgartenschau展馆是一个原型建筑，展现了正在进行中的轻型木结构新型建筑可能性的研究。该项目采用了计算设计及机器人建造方法，成为首个主结构由机器人预制的山毛榉胶合板构成的单元式木板壳体结构。展馆壳体厚度仅为50mm，是一个具有突出性能，资源高效利用的结构。结构采用了当地可再生木材，成为从社会和经济角度综合考虑的可持续发展建筑的典范。

20世纪建筑和材料科学的发展带来了钢材、混凝土及复合材料，木材这一最古老的建筑材料，乍看之下似乎与创新结构相违背。但受到新型性能化木制品发展的影响，木材正重新受到青睐。此外，计算机数控机器尤其是工业机械臂为建造带来的无穷可能性与在建筑行业中的日益增长的自动化目标相契合。最后，建筑行业作为气候变化的主要贡献者，正面临诸多挑战，我们更清晰地认识到：可持续的当地木材的使用可大大降低施工对环境的影响。

The Landesgartenschau Exhibition Hall is a building prototype that demonstrates on-going research into new architectural opportunities for lightweight timber construction. Enabled by integrative computational design and robotic fabrication methods, it is the first segmented timber plate shell to have its primary structure robotically pre-fabricated from beech plywood. With a shell thickness of only 50mm, the building is an extraordinarily performative and resource efficient structure. Being made from the locally available and renewable resource wood, it is also an example for a socially and economically integrated sustainable architecture.

Given the advances in building and material science in the 20th century with the advent of steel, concrete and composites, utilizing wood, which is one of the oldest building materials, in innovative construction seems at first glance antithetical.

作为"机器人木构建造"研究项目的一部分，Landesgartenschau展馆旨在展现新型的、机器人建造技术如何结合计算机设计、模拟和各种测量方式，为建筑木材开拓全新的设计机遇及应用领域。该项目由欧洲区域发展基金（ERDF）及巴登-符腾堡州政府部分赞助，在斯图加特大学及木材加工公司——Müllerblaustein Holzbau GmbH公司的建筑及工程研究人员的紧密协作下完成。该项目的重点是将建筑实体与建造规则整合到轻型的咬合节点木板结构中去。

Landesgartenschau展馆在至少以下五个不同的领域引入了创新设计：① 仿生轻量化设计，将海胆的骨架转化成性能化生形法则；② 通过新颖的基于代理的建模及模拟方式，在建筑设计阶段将建造、安装及装配限制条件与设计流程相整合；③ 在对于单元构件的预制时使用了精准度较高的7轴机器人铣削；④ 凭借先进的测量方式保证质量；⑤ 采用新颖的建筑综合方式，打造出高效的高度可持续结构，壳体相对厚度仅为1/200。

该项目的山毛榉胶合板壳体包含了243个独特的木板单元构件以及7 600个独立的咬合接头，这些接头在结构及建筑中发挥着重要作用。展馆确保了木板边缘四周可以进行高效荷载的转换；同时，可从内部清晰地观察到这些接头构造，打造出独特的建筑体验。建筑外型与木板几何形状的关系在穹顶区及鞍形区的转换中逐渐明显——木板逐渐由凸面转变为凹面。最终，Landesgartenschau展馆不仅是新型木结构系统的范例，还成为了一种新颖的极具表现性的建筑。

However, wood is currently experiencing a resurgence of interest due to the development of new and performative wood products. Additionally, its high fabricability using Computer-Numerically Controlled machinery and in particular industrial robot arms is in line with the increasing automation of the building industry. Lastly, given the current challenges that the building industry is facing as a main contributor to climate change, it is becoming increasingly clear that sustainably and locally sourced wood has the potential to significantly reduce the environmental impact of construction.

As part of the research project "Robotics in Timber Construction", the aim of the Landesgartenschau Exhibition Hall is to demonstrate how novel robotic fabrication processes in conjunction with computational design, simulation, and surveying methods, offer entirely new design possibilities and fields of application for wood in architecture. Partially funded by the European Regional Development Fund (ERDF) and the state of Baden-Württemberg, the project was conceived in close collaboration between architecture and engineering researchers at the University of Stuttgart and the timber construction company Müllerblaustein Holzbau GmbH. A particular focus was the further development and integration of the requirements of building physics and building code into a lightweight finger-joint plate structure.

The development of the Landesgartenschau Exhibition Hall introduces design innovation in at least five different areas: ① biologically informed lightweight design by transferring performative morphological principles of the plate skeleton of sea urchins; ② architectural design methodology by integrating fabrication, construction, and assembly constraints early into the design process through novel agent-based modelling and simulation methods; ③ highly accurate 7-axis robotic milling during pre-fabrication, which has been validated by ④ advanced surveying methods for quality assurance; and finally ⑤ a novel architectural synthesis resulting in a highly sustainable and resource efficient structure with a relative shell thickness of only 1/200.

The resulting beech plywood plate shell consists of 243 geometrically unique plates with a total of 7600 individual finger joints, which play an important structural and architectural role. While ensuring an efficient load transfer around the plate's edges, the joints are also visible on the interior and are part of the architectural experience. The relation between the building's shape and the geometry of the plates becomes apparent in the transition between the dome-shaped and saddle-shaped zones where the plate outline gradually changes from convex to concave. Finally, the Landesgartenschau Exhibition Hall is not only an example of a new type of timber construction system but also a novel and expressive architecture.

由轻质的山毛榉胶合板构成的Landesgartenschau展馆外壳结构安装。
Assembly of the structural shell of the "Landesgartenschau Exhibition Hall" composed of the lightweight beech plywood segments.

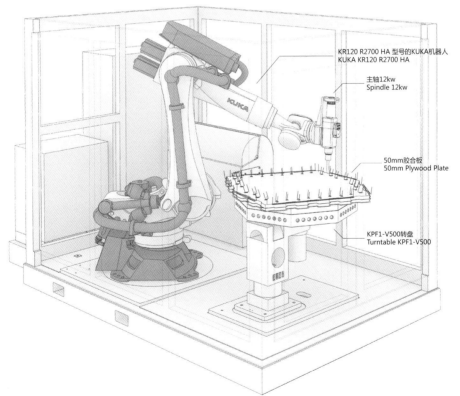

KR120 R2700 HA 型号的KUKA机器人
KUKA KR120 R2700 HA

主轴12kw
Spindle 12kw

50mm胶合板
50mm Plywood Plate

KPF1-V500转盘
Turntable KPF1-V500

机器人建造首先是要模拟建造过程和生成机器人控制的代码，再进行铣削加工，将山毛榉胶合板加工成有大尺度咬合接头的板材，这样能按照现场的要求并符合建筑规范的预制板材就加工完毕了。

The robotic fabrication starts with the simulation of fabrication process and the generation of robot control code. Then follows the milling process: robotic fabrication of a beech plywood plate with large-scale finger joints. After that, pre-fabricated plates that incorporate requirements of on-site assembly and building regulations are complete.

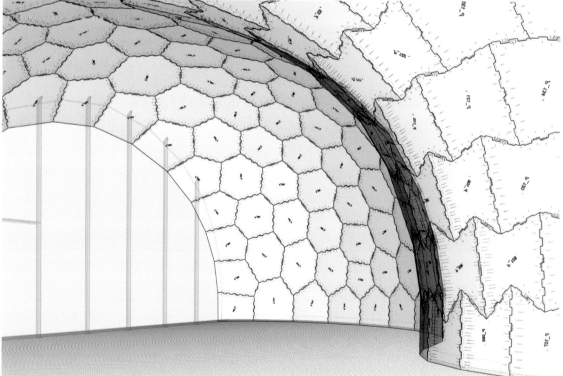

这项目证明了运用当地可用资源进行建筑与结构一体化计算设计和机器人加工的潜力，由计算生成的建筑信息和加工数据模型帮助实现了高精度的加工和标准化的建造。
The building prototype demonstrates the architectural and structural potentials of integrative design computation and robotic fabrication using locally available resources. The computationally generated building information and fabrication data model helps to realize the high-precised and standardized construction.

凹凸的多边形轮廓板材的变化对应了从圆顶过渡到马鞍形曲面曲率。
The change from convex to concave polygonal outlines of the plates corresponds to the transition from dome- to saddle-shaped surface curvature.

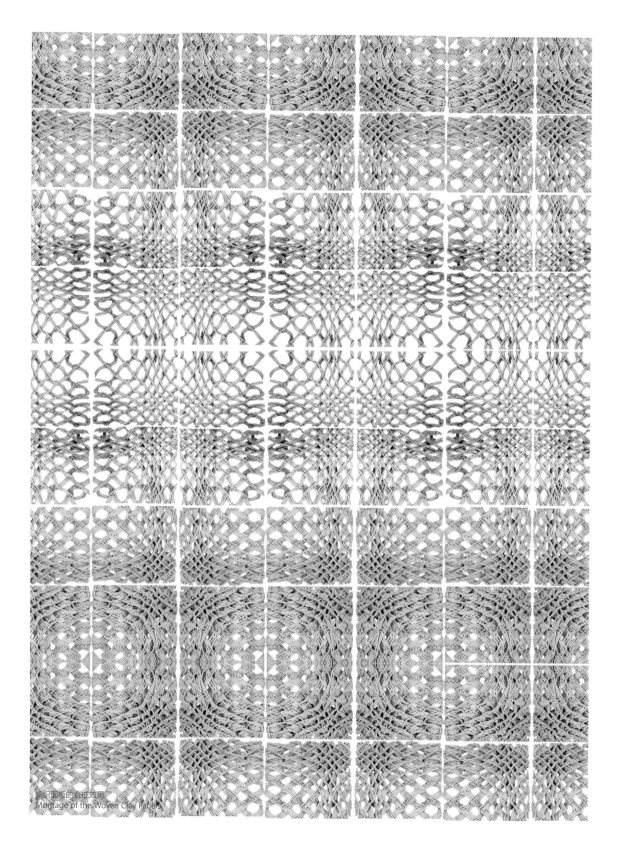

编织面板的合成效果
Montage of the Woven Clay Paper

Woven Clay
Experiments in Robotic Clay Deposition

编织粘土
在机器人粘土沉淀上的实验

Jared Friedman, Heamin Kim, Olga Mesa / Harvard Graduate School of Design
贾瑞德•弗莱德曼，赫敏•金，奥尔加•麦莎 / 哈佛大学设计研究生院

与光线、声音和气味过滤器一样，陶土制品作为建筑中的正规表现形式已长达几个世纪。其中一项最古老、最常见的制陶技术涉及粘土卷的装配工作，这是一项对时间和精确度要求较高的工艺 [1]。有时这些粘土卷会被装配成编织图案，但多数情况下，黏土卷被逐层装配，围合成碗或类似容器的外表面。这项工艺与设计行业盛行的3D打印机的通用方法有诸多相似之处。由于对标准文件类型及沉淀技术的依赖性，3D打印及数字化工作流程的局限性已经凸显。

美国试验材料协会（ASTM）在2012年将"增材制造"定义为"材料粘合工艺，通常使用逐层堆叠、累积的方法，利用3D模型数据构造物体" [2]。"编织粘土"项目通过扩充常用打印技术词汇的方法挑战该定义，应用了将各层材料编织到一起的沉淀技术 [3]。通过在材料挤压中应用传统粘土卷，编织技术打造出用常规"逐层累积"打印技术无法实现的图案及形状。这些图案及形状在平衡6轴工业机器人设定的控制、尺寸及速度的同时，还考虑了材料的性质。工业机器人黏土沉淀工艺对陶土面

Ceramics have been used for centuries in architectural applications as a means of formal expression as well as a filter of light, sound, and smell. One of the oldest and most universal ceramic techniques involves the assembly of clay coils – a process which is time-demanding and requires great precision [1]. While sometimes these coils are assembled in woven patterns, most often they are assembled layer-by-layer in order to build up the surface of a bowl or similar vessel. Such a process shares many similarities with the common approach to 3D-printing machines that have become pervasive in the design industry. The machines and digital workflows common to the field of 3D-printing have been somewhat limiting due to their reliance on standard file types and deposition techniques.

ASTM (2012) defines "additive manufacturing" as the "process of joining materials to make objects from 3D model data, usually layer upon layer." [2] "Woven Clay" challenges this definition by expanding the vocabulary of the printing techniques commonly

板构造的影响是本研究的重要部分。这与戈特弗里德·森佩尔描述的机器制造在工业革命时期对纺织品风格的影响有相似之处。森佩尔主要强调有形机器影响,当前做法则要求我们更多地考虑数字化领域向实体转化的影响。因此,风格不仅受到机器人沉淀工艺影响,也受到机器人所需数字化工具的影响[4]。

早期开发的原型采用了人工粘土挤压方法,同时也考虑到了工业机器人的局限性和潜力。在深刻理解材料和人工方法局限的基础上,建造工艺被转化成了预期的工业机器人设定模式。通过使用数字工具Grasshopper及瑟伯特·施瓦兹开发的HAL机器人编程及控制插件,生成了机器人刀具轨迹及RAPID代码。这些工具允许高度定制及自动化工作流程,可根据输入的几何图形自动更新。此外,研究采用了机器人粘土沉淀策略作为建造编织建筑面板的方式,这种面板外形受光线调整和过滤所驱动。与人工方法相比,粘土沉淀工艺制造速度更快,型材更加统一。

机器人工具及其移动方式

粘土挤压末端执行器的开发采用了早先哈佛大学设计学研究生院设计机器人小组编制的工具,并做了微小改动。这些变动中最值得一提的是挤压机端头的开发,这个端头可在直径为3/8in时挤压粘土卷(直径可根据所需分辨率及打印时的沉淀速度确定)。挤压机通过机械装置运行,设有齿轮马达将导螺杆驱动进入活塞,活塞将黏土推到定制喷嘴上[5]。面板尺寸(45.72cm×45.72cm)主要取决于筒上可以装载的粘土数量。筒的尺寸则根据工具与机器人的相对比例及机器人的有效荷载确定。对于每个面板来说,采用手动方法向筒中填充粘土,筒中剩余的粘土可在下次面板制造中重新利用。每个面板所需平均机器运行时间约为8min。由于需要将该工艺设定于工业环境下,且机器运行时间对于工艺的经济可行性至关重要,因此需密切进行监控。受限于筑造一栋建筑物规模的产品所需机器运行时间的经济性,现在许多3D打印无法成倍放大。

除机器人工具和面板比例外,众多检测材料同机器人动作及旋转速度一致性的测试非常必要,以确保粘土沉淀的准确性。由Grasshopper制作的B曲线被用作刀具轨迹的中心线,这些曲线必须设置在面板的打印平台上方一定距离,以确保沉淀的准确度和一致性。沉淀性质同时也取决于刀具轨迹的弯曲度,这主要是由于机器人在跟随刀具轨迹时会加速和减速。为解决这个问题,有必要利用Grasshopper计算刀具轨迹的弯曲度并用作机器人速度的乘数。这确保了机器人可以在高曲率的区域加速移动,在低曲率的区域减速移动。

一旦根据美观和孔隙水平确定了理想的造型,就可将特定逻辑运用于打印路径。一层黏土沿着模具底座周边路径连续沉淀,可缩减用于停止和切割黏土的时间,同时也确保粘土可粘在模具四边,不至于滑落或移位。边缘上切下来的多余粘土最终还可重新利用。

used in order to incorporate deposition techniques that begin to weave the layers of material together [3]. Using the traditional clay coil as the material extrusion, the weaving technique allows for the creation of patterns and geometries that do not exist in the common "layer upon layer" printing techniques. The creation of such patterns and geometries considers the nature of the material while also leveraging the control, size, and speed granted by the use of a 6-axis industrial robot. The impact of the industrial robotic clay deposition process on the tectonics of the ceramic panels was a critical part of the investigation. This interest draws many similarities to the ways in which Gottfried Semper describes the impact of machine fabrication on the styles of textiles during the industrial revolution. While Semper spoke primarily on the impact of the physical machine, current practice requires us to also consider the impact of the translation from the digital realm to the physical. Thus not only is style impacted by the robotic deposition process, but also by the digital tools that are necessitated by the robot [4].

Early prototypes developed utilized manual methods of clay extrusion while keeping in mind the limitations and potentials of the industrial robot. With a better understanding of the material and limitations of the manual methods, the fabrication process was translated to the intended industrial robot setup. Toolpaths and RAPID code for the robot were generated using the digital tools Grasshopper and the plugin HAL Robot Programming & Control developed by Thibault Schwartz. These tools allowed for a highly customizable and automated workflow that would automatically update based on the input geometry. Additionally, the research utilized a robotic clay deposition strategy as a means of fabricating woven architectural panels whose forms were driven by the modulation and filtration of light. This clay deposition process enabled quicker and more consistent extrusions than were achievable with manual methods.

Robotic Tooling and Movements

Development of the clay extrusion end effector utilized a tool formerly developed by the Harvard GSD Design Robotics Group, in which some minor alterations were made. Most notable of these adjustments was the development of a tip for the extruder that would extrude coils at a 3/8" diameter, which was determined based on the desired resolution and deposition speed that it allowed for while printing. The extruder operates via mechanical means with a gear motor that drives a lead screw into a plunger that pushes the clay through a custom nozzle [5] The size of the panels (45.72cm×45.72cm) was based largely on the amount of clay that the canister on the tool can hold. The size of the canister was based on the relative scale of the tool to the robot, as well as the payload of the robot. For each panel, the canister was

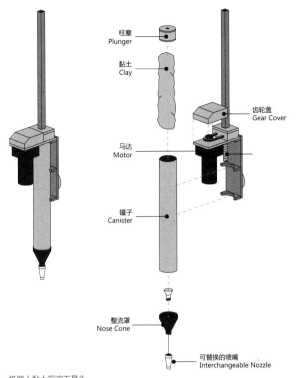

机器人黏土沉淀工具头
Robotic Tool of Clay Deposition

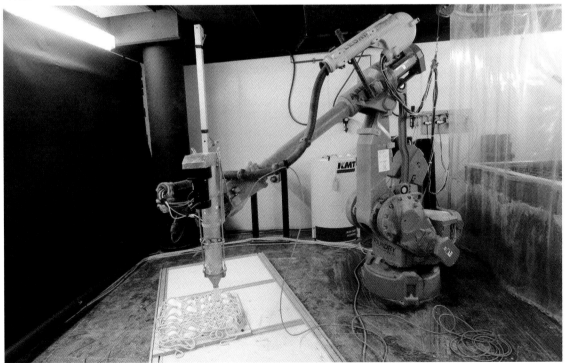

在研磨的印床上做机器人黏土沉淀。
Robotic clay deposition over milled printing bed.

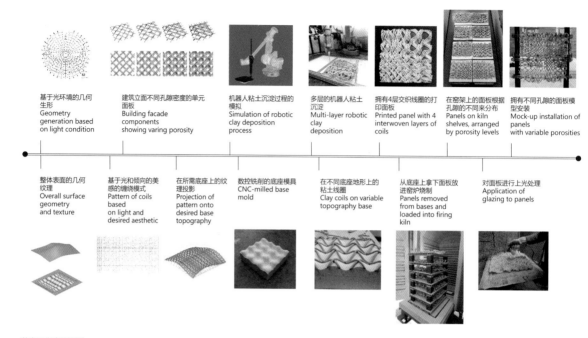

数字工作流程图解
Diagram of Digital Workflow

数字工作流程

数字工作流程始于开发用作打印基础的表面形态，紧接着就是将B曲线转化为工具轨迹，并发送给机器人。本项研究的目标之一就是证明可以通过改变打印基础的形态达到立体效果，将平面形状改造为更具变化的形态。因此，数字流程的第一步就是形成可用于机器人粘土沉淀的基础模具的曲面。利用正弦曲线发展凸起的表面，以便强调深度上的变化。正弦曲线遵循间距不同的网格，以便表达各面板之间的密度变化梯度。Grasshopper的参数工作环境方便轻松复制网格以及凸起的深度。面板的总体深度变化非常细微，但却证明了制作更加极端的几何曲线的可能性。

一旦绘制了曲线形态，就可利用正弦曲线和相同的曲线网格形成图案。在这种情况下，利用Grasshopper绘制曲线，并同最初的网格对角连接。接着利用Grasshopper和HAL组件将这些曲线转化成刀具轨迹。虽然形成刀具轨迹的程序有数个，但HAL的优点在于允许数字工作流程保持在Rhino 和Grasshopper的环境中。这意味着基础曲面或曲线设计的任何扭动都将自动更新刀具轨迹，并将编码迅速发送给机器人。该流程的优势在于从设计到加工流程极为迅速，便于迅速定制化面板。就商业生产规模而言，工作流程可方便设计师选择诸如屏幕透明度等，并紧接着自动形成几何形状和刀具轨迹的过程，避免直接调整输入的曲面和曲线，而是直接关注屏幕上的预期效果即可。

manually refilled with clay, and any leftover clay in the canister was able to be reused for the next panel. The average panel took approximately 8 minutes of machine time. This was something that was monitored closely since in order to propose this process in an industrial setting, machine times are crucial to the economic viability of the process. Many current explorations in 3D-printing fail to scale up due to the machine times that are required in order to construct something at the scale of a building.

In addition to the robotic tooling and scaling of the panels, many tests with the consistency of the material and the speed and rotation of the robotic movements were necessary in order to ensure an accurate clay deposition. B-splines generated in Grasshopper were used as the centerline of the toolpaths. These splines had to be offset at a specific height from the printing bed in order to ensure the most accurate and consistent deposition. The behavior of the clay deposition also depended on the degree of curvature of the toolpath. This is due primarily to the acceleration and deceleration of the robot as it follows the toolpath. In order to resolve this it was necessary to calculate the degree of curvature of the toolpath in Grasshopper and use that as a multiplier for the speed of the robot. This then ensured that the robot would move more rapidly over areas of high curvature and more slowly over areas of low curvature.

MakerBox之外的思考

本研究表明了超越一般的增量制作做法的潜力。近年来增量制作快速发展，进一步增加了机器和软件的经济性和灵活性。但由于文件格式和数字模型落后，仍然有许多限制。例如，基于网格的STL文件格式最近几十年来一直被视作3D打印的标准。该格式将一个曲面切割成数层，打印时依照一个曲面上叠加另一个曲面进行打印。由于工业机器人的设计与这种工作流程不兼容，刀具轨迹的设计将3D打印从层层叠加的方式中解放出来。可以看到，工业机器人最大的资产就是不再受限于自计算机辅助设计有史以来所逐渐形成的通用的业界标准。

本义所述的机器人粘土沉淀在更大范围内的应用很大程度上取决于制造的速度与可生产的面板尺寸的关系。如果我们把该工艺比作建筑小组件的3D打印技术，我们便可以发现更高效、更经济的增量制造工艺。通过本研究研发的原型尺寸受筒装粘土数量和烧制土窑尺寸的限制，但在工业环境下，这些参数的限制性可能比较小。在更大的范围内，这些限制更有可能源自陶土原料的结构特性，同时也取决于所采用的装配方式。本研究的首要目标是探索机器人沉淀工艺的新机遇。生成的原型仍有深化的空间，它们是高效的自动设计流程的体现，通过挑战增量制造的一般定义可以实现。

Once a desired pattern was determined based on aesthetics and porosity levels, a specific logic was applied to the printing paths in order to maximize overlaps and interweaving - thereby decreasing the fragility of the panels. A single layer of coil is deposited along one continuous path that overshoots the edges of the base mold. This is done to reduce time spent stopping and cutting coils, while also allowing the clay coils to catch the edges of the mold so that they don't slide or shift. All excess clay cut off of the edges at the end is able to be reused.

Digital Workflow

The digital workflow began with developing a surface topography to be used as the printing base, followed by the formation of the B-splines to be converted to toolpaths that were then sent to the robot. One of the goals of the research was to provide evidence that dimensionality can be achieved by altering the topography of the printing base from one that is flat to something with more variation. Thus, the first step of the digital process involved generating surfaces to be used as the base molds for robotic clay deposition. Bumped surfaces were developed based on sine curves in order to emphasize a variation in depth. The sine curves followed a grid with variable spacing in order to express

拥有不同孔隙的面板模型安装
Mock-up installation of panels with variable porosities

gradients of densities across each panel. The parametric work environment of Grasshopper allowed for easily manipulation of the grid, as well as manipulation in the depth of the bumps. While the overall depth of the panel only had subtle shifts in its dimensions, it provided a proof-of-concept that more extreme surface geometries are possible.

Once the surface topography was developed, spline curves were used in the formation of a pattern that utilized the same grid in which the surface was developed. In this case, splines were developed in Grasshopper that connected diagonally along the initial grid. These splines were then translated into toolpaths using Grasshopper and HAL components. While there are a number of programs that could have been used to generate toolpaths, HAL allowed the digital workflow to remain within the Rhino and Grasshopper environment. This meant that any tweaks in the design of the base surface or splines would automatically update the toolpaths and RAPID code to be sent to the robot. This process allowed for a quick design to fabrication workflow that allowed for quick customization of the panels. At the scale of commercial production, a workflow might be considered that would allow the designer to select features such as the overall opacity percentage of a screen, which would then automatically generate the geometry and toolpaths. This would avoid having to directly adjust the input surfaces and splines, and instead focus on the desired effects of the screen.

Thinking Outside the MakerBox

The research performed throughout this project displays the potential to expand beyond commonly accepted practices within additive manufacturing. Rapid development in the additive manufacturing industry in recent years has led to further affordability and flexibility of both machines and software. However, many of the limitations that remain are due to outdated file formats and digital modeling practices. For example, the mesh-based STL file format has been viewed as the standard in 3D-printing for decades. This format forces the automated slicing of a surface into layers that are then printed one on top of the other. Because the industrial robot was not designed to accommodate this type of workflow, the designing of the toolpaths was freed from common limitations imparted by machines that assume a layer upon layer approach. Here we see that one of the greatest assets of the industrial robot in this context is its ignorance to the de-facto design practices that have developed within in the field since the early days of computer aided design.

The potential for the robotic clay deposition process outlined here to be applied at a larger scale is based largely upon the speed of fabrication in relation to the size of the panel that can be produced. If we compare this process to the 3D-printing of smaller building components such as bricks, we see here a more efficient and economical additive manufacturing process. While the size of the prototypes that were developed throughout this research were limited based on the amount of clay the canister can hold and the dimensions of the firing kilns used, in an industrial setting these parameters could be much less limiting. At a larger scale, limitations are more likely to come from structural properties of the ceramic material, which would also depend on the assembly methods employed. The primary goal of the research was to explore new opportunities granted by the robotic deposition process. While the resulting prototypes leave much room for development, they are a testament to the efficiency of automated design processes, and the formal potential that can be achieved by challenging the common definitions of additive manufacturing.

参考文献 / References：

[1] Peterson, S., Peterson, J. 2003, Craft & the Art of Clay: A Complete Potter's Handbook: Fourth Edition, Lawrence King Publishing Ltd. – London, UK, pp.33-34.

[2] ASTM F2792-12a Standard Terminology for Additive Manufacturing Technologies 2012, ASTM International, West Conshohocken, PA.

[3] Friedman, J., Kim, H., Mesa, O., 2014 'Experiments in Additive Clay Depositions.' Robotic Fabrication in Architecture, Art, and Design 2014, Springer International, Switzerland.

[4] Semper, G 1989, The Four Elements of Architecture and Other Writings, Cambridge University Press, Cambridge.

[5] King, N., Bechthold, M., Kane, A., 2012 'Customizing Ceramics: Automation Strategies for Robotic Fabrication.' Fabricating the Future, Tongji University Press.

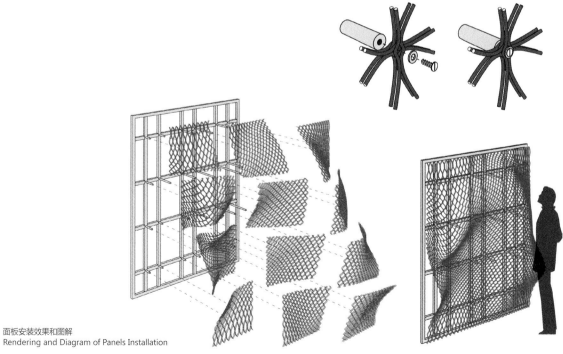

面板安装效果和图解
Rendering and Diagram of Panels Installation

2014 "数字未来"暑期学生设计工作营中的"反转檐椽"项目
Reverse Rafter, DigitalFUTURE Shanghai 2014 Summer Workshop

Reverse Rafter
Structural Performance Simulation Based On Wood Tectonics
反转檐椽
传统木构的结构性能模拟

Philip F. Yuan, Chai Hua / CAUP, Tongji University
袁烽，柴华 / 同济大学建筑与城市规划学院

随着数字建造技术的快速发展，结构性能化设计展现出广阔的应用前景。本文基于研究项目"反转檐椽"，旨在探索基于结构性能的数字技术的辅助下模拟与再应用中国传统木构的可能性。项目以"檐椽"为研究原型，采用拓扑优化的研究方法，引入"千足虫"（millipede）为分析工具。通过传统结构计算方法和拓扑优化方式的结果比对，项目揭示了"檐椽"潜在的结构原则。在此基础上利用数字方法设计了一个现代互承结构装置。在数字建造过程中，项目采用CNC切割技术来确保建造精度。

With the rapid development of digital fabrication technologies, structural performance based design has a broad range of potential applications. Based on the research project, "Reverse Rafter", this paper aims to explore the possibility of simulating and re-applying Chinese traditional wood tectonics with structural performance based computational technology. Taking the "eaves rafter" as a research prototype, this project employs topological optimization as a research method and "Millipede" as an analysis tool. Through a comparison between the results of traditional structure analysis and the topological optimization method, this project reveals the underlying structural principles of the "eaves rafter". On the basis of this a modern reciprocal structure installation was generated through a digital design method. CNC cutting technologies were employed to ensure accuracy in the digital fabrication process.

基于结构性能的设计方法是一种通过结构性能模拟、计算和优化寻找空间形式和结构之间合理关系的设计过程。不同于图解静力学时代（19世纪末 – 20世纪中）、力学建构时代（20世纪中 – 20世纪末）的结构设计方法，数字生形时代（20世纪90年代 – 至今）的结构性能化设计关注于结构分析与设计之间的动态与交互，使得基于结构性能的多目标、多维度的建筑形态生成成为可能 [1]。

文化是技术发展的基本语境。木构作为中国传统建筑文化载体，在中国建筑文化的当代实践和研究中扮演着重要角色。结

Structural performance-based design methodology is a design process to find a reasonable relationship between spatial form and structure through simulation, calculation and optimization of structural behavior. Unlike the structural design methods of the Graphic Statics Era (late 1800s – mid 1900s) and the Structural Tectonics Era (mid 1900s – late1900s), structure performance-based architectural design in the Digital Morphologies Era (1990s –) focuses on the dynamic interaction between structure analysis and design, enabling the multi-objective and multi-dimensional morphological analysis and optimization based on structural behavior [1].

构性能化设计技术能够将技术与文化相结合，为建筑师科学地处理传统建筑工艺提供了一套解决方案。

20世纪90年代，数字技术的出现带来了大量结构性能化计算方法和分析工具，为数字生形时代的结构性能化设计提供了有力支持。本文通过对传统结构原型的分析，试图探究数字模拟与优化中国传统木构系统的可能性。

分析方法和工具

基于有限元的分析，新兴结构分析方法拓扑优化法通过将结构体细分为有限个离散的单元，基于某种算法决定结构材料的去留。拓扑优化在寻找一定约束条件下结构拓扑形式和尺寸方面具有巨大优势。拓扑优化为找形提供了合理的力学基础和结构参照，逐渐被建筑师和工程师所认可。随着计算技术的高速发展，大量基于拓扑优化的算法或程序包如ESO（渐进结构优化）和BESO（双向渐进结构优化）被开发出来，为拓扑优化的实践应用提供了重要基础[2]。

在这些拓扑优化工具中，帕纳约蒂斯·米哈拉托斯基于犀牛软件开发的插件"千足虫"以其性能优势成为新一代结构优化工具的代表。这一插件包括一个快速线性弹性系统结构分析算法库，能够进行平面力作用下框架的快速弹性线性分析，将分析结果提取并以多种方式被可视化[3]。

中国传统木构以线性杆件元素为主，"千足虫"的拓扑尺寸优化功能为木构的结构模拟与优化提供了可能和基础。

"反转檐椽"是"2014上海数字未来"①活动的一个结构装置。该项目以"千足虫"为模拟工具，通过设计与建造实践为中国传统木构的数字模拟与优化提供了新的可能。

原型研究

出檐深远的坡屋顶是中国传统建筑最主要的形式特征，呈现出优雅的美学表现和合理的结构受力。檐椽对于支撑悬挑的屋面起着至关重要的作用，该项目以"檐椽"为研究原型。

传统屋顶主要以两端置于檩上的椽承受自重、雪等荷载，屋檐部位的出挑对结构性能提出了很高的要求。檐椽由正心桁、老檐桁以及由斗拱承托的挑檐桁共同支撑，斜向下方向悬挑而出，以承托上部屋面或飞檐。

《清式营造则例》是从《清工部工程做法则例》②中提取的清代官方设计规程和营造标准。书中规定檐椽的出挑尺寸——从挑檐桁至檐口的水平距离——为14斗口③，挑檐桁到正心桁的水平距离为6斗口，正心桁到老檐桁为24斗口（见右页表1和表2）[4]。但书中并未给出明确的依据，并不能确定檐椽的出挑尺寸是结构因素使然，还是另有考虑。国内学者曾经将檐椽简化为一次超静

Culture provides the context for the development of technologies. As a traditional architectural material, wood plays an important role in contemporary practices and research in Chinese building culture. The development of structural performance-based technologies offers a way for architects to deal with traditional tectonics scientifically, integrating technology with culture.

The advent of digital technologies in the 1990s has produced many structural performance-based computing methods and analysis tools, which provide powerful support for the structural performance-based design in the Digital Morphologies Era. Through the analysis of the traditional structural prototype, this article aims to explore the possibility of digital simulation and optimization for Chinese traditional wood tectonic systems.

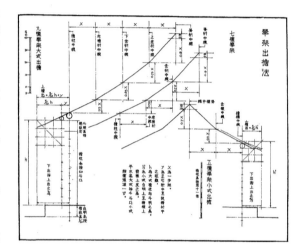

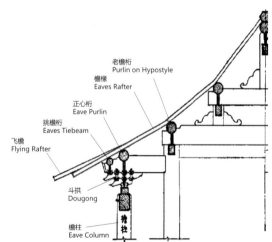

檐椽由正心桁、老檐桁以及由斗拱承托的挑檐桁共同支撑，斜向下方向悬挑而出，承托上部屋面或飞檐。《清工部工程做法则例》中规定了檐椽的出挑尺寸。
Supported by eaves purlin, purlin on hypostyle and eaves tiebeam, eaves rafter cantilevers out obliquely to support the overhanging part of the roof or flying eaves. The book "Qing Gongbu Gongcheng Zuofa Zeli", provides the overhanging size of eaves rafter.

表 1 檐椽与飞檐椽的尺寸关系
Table 1 Size relationship between Eaves rafter and Flying rafter

	大木大式 Wooden frame with dougong	
	高度或长度 Height or length	截面 Cross-section
飞檐椽 Flying rafter	3/10*(柱高+斗拱高度) 3/10*(column height+dougong height)	1.5斗口 1.5doukou
檐椽 Eaves rafter	3/5*飞檐椽长度 3/5*flying rafter length	1.5斗口 1.5doukou

表 2 檐椽相关数据
Table 2 Relevant data about eaves rafter

柱 Column	大木大式 Wooden frame with dougong	
	高度 Height	截面 Cross-section
檐柱 Flying rafter	60斗口 60doukou	6斗口(1/1000收分) 6doukou(1/1000contracture)

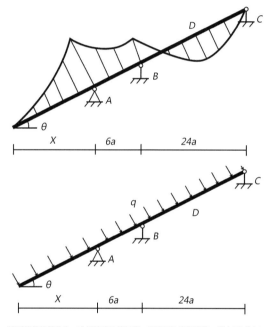

通过将檐椽简化为一次超静定连续斜梁，平衡正负弯矩峰值，得出X约为14a。结构计算简化模型（上）；弯矩图（下）。
By simplifying the eaves rafter into a statically indeterminate oblique beam and balancing the positive and negative peak moments, the X come out to be 14a. Simplified rafter model for calculation (up); Bending moment diagram (down).

Analysis Methods and Tools

Based on the finite element analysis method, topology optimization – the emerging structural analysis method – determines the reduction or retention of structural material by dividing the structural volume into discrete finite units according to a specific algorithm, thereby proving itself highly effective in finding the optimal topological shape and size of structures under certain constraints. Topology optimization provides reasonable structural foundation and reference for form finding, as is gradually being recognized by architects and engineers. With the rapid development of computational technologies, many algorithms or programs based on topology optimization have emerged such as ESO (Evolutionary Structural Optimization) and BESO (Bi-directional ESO), which provide an important foundation for the practical application of topology optimization [2].

Among topology optimization tools, the Rhino based plug-in "Millipede" developed by Panagiotis Michalatos has become a popular structural optimization tool due to its efficiency. This software comprises a library of fast structural analysis algorithms based on topology optimization for linear elastic systems. It allows for a very fast linear elastic analysis of frame in plane forces, having their results extracted and visualized in a variety of ways [3].

The topology sizing optimization function of "Millipede" provides

---

① 上海"数字未来"暑期工作营是始于2011年的系列学术活动。该活动每年有一个主题。2014年"数字未来"活动的主题为"基于结构性能的机器人建造"。
DigitalFUTURE Shanghai is a series of academic events starts from 2011. Every year it will have a topic. "Robotic Fabrication Based on Structural Performance" is the topic of DigitalFUTURE 2014.

② 《清部工程做法则例》是清朝营造规程，相当于宋朝《营造法式》。
"Qing Gongbu Gongcheng Zuofa Zeli" is the technical treatise in Qing Dynasty of China, similar to "Yingzao Fashi" of Song Dynasty.

③ 斗口，清朝模数单位。
Doukou, modular system in Qing Dynasty.

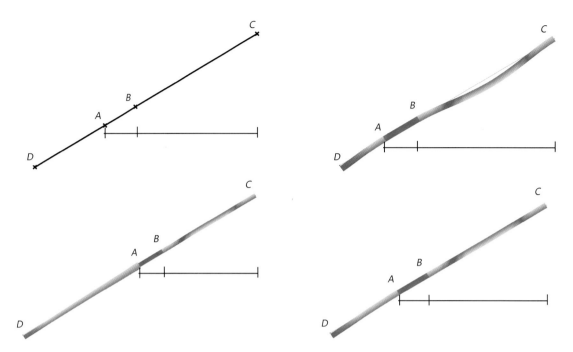

檐椽结构模拟模型（左上）；AD长度与AC长度比例为0.2时的平衡状态（右上）；AD长度与AC长度比例为0.8时的平衡状态（左下）；AD长度与AC长度比例为0.47时的平衡状态（右下）。
Simplified rafter model for simulation (up left); Equilibrium state when AD/AC equals 0.2(up right); Equilibrium state when AD/AC equals 0.8(down left); Equilibrium state when AD/AC equals 0.47(down right).

定连续斜梁[5]，运用结构计算和弯矩图进行分析，通过平衡正负弯矩峰值，得出了与《清式营造则例》的描述基本一致的结果。

结构计算方法证实了檐椽出挑尺寸规定的结构理性，但并没有真正揭示其中隐含的结构原理。与之相比，数字分析软件能够实现更加动态的分析和优化。该项目用"千足虫"模拟了"檐椽"的受力环境。首先在软件中建立一个包含材料信息、受力状态、外部约束的分析模型。在模拟过程中，详细观察结构形态随着出挑比例的改变所发生的变化。模拟结果表明：在相同的杆件斜率下，悬挑比例的减小会引起支撑点B、C之间的变形增大；悬挑增大时，A点需要加大杆件尺寸才能抵抗弯矩的作用。因此适当的悬挑比例应该是权衡BC段的变换A点的位置的结果。经过多次尝试得出，当AD段与AC段的比例在0.46~0.48之间时，能够获得上述平衡。

进一步研究发现，当支撑点B的位置改变时，AD段与AC段的均衡比例并不受影响，始终维持在0.46~0.48之间。上述研究与《清式营造则例》中的描述基本一致。研究充分证明多跨连续梁的出挑比例存在一个最合理取值。在多跨连续梁的实践中应根据具体情况采用合理的出挑尺寸，以节约材料、节省预算。与传统结构计算方法相比，"千足虫"的优势在于其动态性互动界面；然而，"千足虫"的分析结果作为一种结构图解，需要进一步解读才能达到实际应用的合理程度。

上述分析充分证明：中国传统营造制度中的规定是中国工匠数

the opportunity and basis for structural simulation and optimization of Chinese traditional wood tectonics which is composed mainly composed of linear struts.

"Reverse Rafter" is a structural installation from the "DigitalFUTURE Shanghai 2014" Summer Workshop [1]. Taking "Millipede" as a simulation tool, this project offers the possibility of computational simulation and optimization of Chinese traditional wood tectonics through the practice of design and fabrication.

Prototype Research

The pitched roof with its deep overhangs is the most prominent feature of traditional Chinese architecture, and presents both an elegant aesthetic expression and a reasonable structural solution. The project took the "eaves rafter" as a prototype that plays the important role of supporting the overhanging structure.

Traditional roofs mainly bear loads such as self-weight and snow loads through rafters laid on purlins. The overhangs at the eaves are required to provide a higher level of structural performance. Supported by the eaves purlins, the purlins on the hypostyle and the eaves tie-beam, the eaves rafters cantilever out obliquely to support the overhanging part of the roof or flying eaves.

The book "Qingshi Yingzao Zeli", which lists the design principles

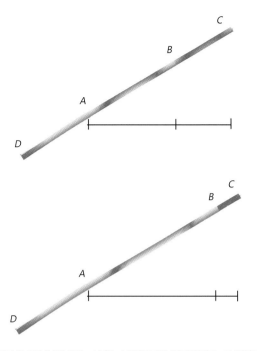

当支撑点B的位置改变时，AD段与AC段的均衡比例并不受影响，始终维持在0.46~0.48之间。AB长度与AC长度比例为0.56时的平衡状态（上）；AB长度与AC长度比例为0.8时的平衡状态（下）。
when the position of support point B was changed, the balanced proportion between AC and AD was almost unaffected, maintaining at 0.46-0.48 range. Equilibrium state when AB/AC equals 0.56 (up) ; Equilibrium state when AB/AC equals 0.8 (down).

千年的经验积累，蕴含着丰富的结构信息。结构分析有助于将结构信息转换成科学理论；同时，分析结果与传统之间的契合也证明了数字化结构工具的有效性和潜力。

结构性能模拟与优化

在原型研究的的基础上，该项目进一步研究了结构性能化工具在结构设计中的作用和意义。

为了进行进一步研究，该项目首先在上述结构原则的基础上完成一个结构装置设计。项目利用"千足虫"在形态优化和尺寸优化方面的拓扑优化功能优势，模拟和优化了结构装置的形式和截面尺寸。

项目在设计过程中引入了互承结构 ④。在这种结构体系中，每根杆件与相邻杆件交接形成两个支点，杆件由支点向两端延伸形成悬挑。互承结构中杆件的主要受弯矩作用，受力情况与檐椽类似。在中国传统建筑中，互承结构往往被用来做拱桥、亭子等构筑物的支撑结构。水平延伸性是互承结构的一大特点，传统的互承结构的应用往往利用这种特性建造一个面状的连续

and manufacturing standards of the Qing Dynasty extracted from "Qing Gongbu Gongcheng Zuofa Zeli" ②, specifies the size of the overhanging eaves rafter, as the horizontal distance from eaves tie-beam to the eaves edge as 14 doukou ③. The horizontal distance from the eaves tie-beam to the eaves purlin is specified as 6 doukou, while that between eaves purlin and purlin on hypostyle is 24 doukou (See table 1, 2) [4]. As the book does not give any clear explanation, it is not sure whether the length of the overhang was dictated by structural considerations. Chinese scholars have given a structural interpretation of the length of the overhang through structural calculations and bending moment diagrams[5]. By simplifying the eaves rafter into a statically indeterminate oblique beam and balancing the positive and negative peak moments, the research comes out with a result consistent with the description of " Qingshi Yingzao Zeli " .

Structural calculations verify the logic of the force distribution of the eaves rafter, but do not really reveal the underlying structural principles. By contrast, digital analysis software allows for a more dynamic and interactive process of analysis and optimization.
In this project, the physical environment for the eaves rafter is simulated in "Millipede". To begin with, a rafter model with material information, load condition, and external constraints is created. During the simulation, shape changes are carefully observed with the adjustment of the ratio of the overhang. Simulations show that, with the same slope of strut, the reduction of the overhang would cause the deformation between the support points B and C to increase. When increasing the ratio of the overhang, the cross-section near point A needs to be strengthened to resist bending moment action. Therefore, the appropriate ratio of the overhang should be a compromise between the deformation of BC and section size of point A. After several attempts, we find out that when the ratio of the length of AD and AC falls within the 0.46~0.48 range, the appropriate ratio of the overhang can be obtained.

Further studies show that, when the position of support point B is changed, the ratio of the lengths between AC and AD is almost unaffected, maintaining the 0.46~0.48 range. These results are basically consistent with the description in "Qingshi Yingzao Zeli". It can be concluded that the size of the overhang of a multi-span continuous beam is reasonable. In the practice, the overhanging size of a multi-span continuous beam should be determined according to the specific conditions, thus saving material and reducing costs. Compared with the traditional structure calculation method, the advantages of "Millipede" lie in its dynamic and interactive interface. However, the results of the "Millipede" analysis require further explanation to make them reasonable for applications.

---

④ 互承结构是一种三维自承重结构系统，其中杆件呈环状相互支撑。
Reciprocal structure is a three-dimensional self-supporting system with rods supported mutually in a circle.

 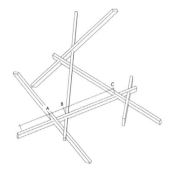 

设计以三根杆件按照合理的悬挑比例互相搭接形成的互承结构为基本单元。图为基于最优比例的单元构件生成过程。
The basic unit was derived from the reciprocal unit with three rods overlapping each other with reasonable overhanging proportion. The figure shows the generate process of a basic unit based on appropriate proportion.

 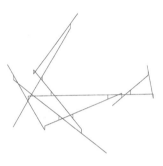 

建模时将每根杆件简化为其中心轴线,在节点处用短线将轴线相连。这种方式使模拟模型中的力流与真实情况相接近。
The simulation model was built with each rod simplified into its central axis with auxiliary line connecting the axis at the joints. In this way, the force flow in simulation model was made closer to real situation.

体。与之相反,这个项目主要利用互承结构进行垂直向生长。

结构原则与特定算法的结合能够产生具有优秀性能的结构。设计以三根杆件按照合理悬挑比例互相搭接形成的互承结构为基本单元。在垂直生长过程中,上层单元在下层单元的基础上生成,因此单元的舒展程度会随着高度的增加而增大,整体呈伞状。在结构设计过程中只需要输入几个基础参数如底层半径,装置的整体形式便会自动生成,同时能够保证每根杆件的性能。

由于每层单元均承受上部所有单元重量,每层单元变形程度因高度而不同。因此,每层杆件需要采用合适的截面尺寸以应对不同大小的作用力。"千足虫"的有限元分析会假定单元之间为完全固定连接,与搭接方式在结构性能上完全不同。因此建模时将每根杆件简化为其中心轴线,在节点处用短线将轴线相连。这种方式使模拟模型中的力流与真实情况相接近。完成模拟模型之后,该项目运用"千足虫"对装置结构性能——轴向力、剪力、弯曲、变形等性能进行了模拟分析。"千足虫"插件自身的拓扑尺寸优化功能能够即时显示优化后的合适的杆件截面,为设计调整提供参照。优化后的杆件截面随受力状态呈现非线性变化。考虑到材料获取的便易程度和加工的经济性,该项目对优化结果进行观察,将材料尺寸归结为数个类别,反

The analysis above fully proves that traditional tectonics accumulated through thousands of years contain a wealth of structural information. Structural analysis nowadays can help us to extract the underlying principles. At the same time, the consistency between the analysis results and traditional standards also demonstrates the effectiveness and potential of digital structural tools.

Structural Performance Simulation and Optimization

On the basis of the analysis of the prototype, the project goes on to study the role and significance of structural performance-based tools in structural design.

As the next step in the research, a structural installation is designed based on structural principles analyzed above. Taking full advantage of the topological optimization function of "Millipede" in shape optimization and material size optimization, this project then simulates and optimizes the form and cross-section of the structural installation.

A reciprocal structure [4] is then introduced in the design phase.

馈到模拟过程中。经过多次循环分析，杆件截面最终被总结为四类：90mm×40mm、70mm×30mm、55mm×20mm、40mm×20mm，优化过程到此为止。在这一阶段，"千足虫"的拓扑优化功能有效地保证了木构的设计与优化。

数字建造

数字生形时代的形式复杂性对建造技术提出了极高的精确性要求。超越传统工艺的力量所及，数字建造技术的发展为结构性能化设计的实现提供了重要支持。

在该项目中，虽然垂直叠加的结构单元遵循同一结构逻辑，但杆件长度、截面尺寸、交接位置、倾斜角度各不相同，从而导致节点的极度复杂。传统手工定位和加工会带来大量误差，而加工误差的累积必然导致整体形式的失控。五轴CNC的精确加工能力能够很好地应对这一需求，使本装置的结构意图得以完美呈现。

该装置共有21种不同尺寸的杆件63根，杆件长度从1~3m不等。加工前，研究团队对每根杆件都进行了编号，由于杆件的独特性，编号对于建造至关重要。节点的定位和切割主要由5轴CNC数控机床加工完成，人工完成现场搭建。与中国传统榫卯结构类似，整个建造过程中没有使用任何连接件、加固构件。

结构最终呈现为由七层互承单元垂直叠加而成的6m高的伞状结构，力流沿杆件呈螺旋状自上而下地传递。装置底部半径只有0.5m，而顶层半径达到3m。装置底部配置的座椅，不仅可以供人休憩，同时作为结构配重，抵抗装置整体倾覆力，保持整体稳定。

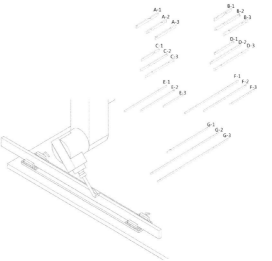

研究团队对每根杆件都进行了编号，节点的定位和切割主要由五轴CNC数控机床加工完成。
The localization and milling process of junctions was mainly performed by the 5 axis CNC.

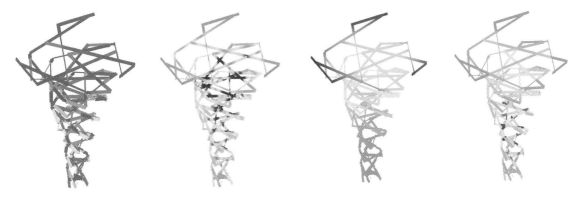

项目运用"千足虫"对装置结构性能——轴向力（1）、剪力（1）、弯曲（3）、变形（4）等性能进行了模拟分析。"千足虫"插件自身的拓扑尺寸优化功能能够即时显示优化后的合适的杆件截面。
he project simulated the structural performance—axial force(1), shear force(2), bending force(3), and resistance to deformation(4) with "Millipede". The built-in topology optimization function of "Millipede" showed the appropriate cross-sections after optimization.

与中国传统榫卯结构类似，整个建造过程中没有使用任何连接件、加固固件。结构最终呈现为由七层互承单元垂直叠加而成的6m高的伞状结构。装置底部半径只有0.5m，而顶层半径达到3m。装置底部配置的座椅抵抗装置整体倾覆力，保持整体稳定。
Similar to Chinese traditional mortise-tenon structure, the entire structure was fabricated without any connecting members or reinforcement components. The final structure presents an umbrella form of 6m high with seven layers. The bottom radius of the structure is 0.5 meters, while the top radius extends up to 3 meters. The triangle bench placed at the bottom counters weight and resist the overturning force to maintain the overall stability.

总结

在数字建造技术快速发展的当下，结构性能化设计拥有广阔的发展前景。本项目将先锋的结构性能设计与传统的木构体系相结合，创造出同时具有文化意义和结构美学的装置作品。这项研究为解读中国传统建筑提供了新的思路，同时传统建筑经验的科学化为建筑结构的发展提供了新的动力和方向。

对中国传统木构的充分研究必将引向一个终极目标，即木材材料性能的研究。然而随着复合木材的发展，木材材料性能本身也在不断发生变化。这就需要我们不断适应材料的演变，充分利用材料特性实现性能化建构。

In the reciprocal structure, each strut overlaps with two adjacent struts, forming overhangs at both ends. Similar to the structural condition of eaves rafters, these struts bear mainly bending moment action. The reciprocal structure often serves as a supporting structure for bridges or pavilions in ancient China. As its horizontal extension is the major feature, the reciprocal structure is conventionally used for building a planar continuum. By contrast, the reciprocal structure developed vertically in this project.

Efficient structures can be obtained by combining the structural principles with certain algorithms. The basic unit is derived from the reciprocal unit with three struts overlapping each other with reasonable overhanging proportions. While the upper unit is generated based on the lower one in terms of its vertical development, the unfolding degree of units increases layer by layer, giving the overall structure an umbrella form. Only a few parameters, such as the radius of the first layer, are needed to generate the overall form and ensure that each individual element behaves adequately.

Since each unit bears the weight of all those above, the degree of deformation varies with height. Therefore cross-sections vary in response to the different force conditions of the different layers. "Millipede" FE analysis assumes totally fixed connections between elements, which display a completely different performance with overlapping connections. Therefore, a simulation model is built with each strut simplified into its central axis with an auxiliary line connecting the axis at the joints. In this way, the force flow in the simulation model is made closer to the real situation. After building the simulation model, the project simulates the structural performance – axial force, shear force, bending force, and resistance to deformation using "Millipede". The built-in topological optimization function of "Millipede" shows the appropriate cross-sections after optimization, which provide clues for design adjustments. Sectional sizes after optimization vary in a nonlinear fashon following changes in the stress state. Taking into account the accessibility and economy of materials, the project observes the optimization results and classifies cross-sections into several types, which then feed back into the simulation. After repeating the analysis loop several times, the cross-sections of the struts are finally divided into four categories: 90mm×40mm, 70mm×30mm, 55mm×20mm, 40mm×20mm, which mark the end to the optimization process. At this stage, the topological optimization function of "Millipede" effectively ensures the design and optimization of wood tectonics.

Digital Fabrication

Form complexity in the Digital Morphologies Era calls for highly accurate fabrication technology. Beyond the capacity of traditional techniques, the rapid development of digital fabrication technologies provides critical support for the realization of comprehensive structural performance-based design.

In this project, although the structural unit of each layer follows the same logic, the strut length, cross-sectional shape, overlapping position and inclination angle vary with each strut, resulting in extremely complex junctions. The accumulation of deviations resulting from traditional manual fabrication processes would inevitably lead to the loss of control of the overall form. The accurate fabricating capacity of a 5 axis CNC milling machine provides a suitable response to this concern, preserving the intention of the design initiatives.

This structure comprises 63 struts with 21 different types of cross-sections, with their lengths varying from 1m to 3m. Before the fabrication process, each strut is labeled in sequence in both digital and physical platforms. As each piece is unique, this was essential for assembling. The localization and milling process of junctions is performed mainly by a 5 axis CNC milling machine, while construction on site is performed manually. Similar to Chinese traditional mortise-tenon structures, the entire structure was fabricated without any connecting members or reinforcing components.

The final structure presents an umbrella form 6m high with seven layers, and with the structural forces passing in a spiral from top to bottom along the struts. The bottom radius of the structure is 0.5 meters, while the top radius extends up to 3 meters. The triangle bench placed at the bottom not only serves as a resting space, but also acts as a structural component as a counter weight to resist the overturning force and maintain the overall stability.

Conclusions

With the rapid development of digital fabrication technologies, structural performance-based design has a broad range of potential applications. This project combines structural performance-based design with traditional wood tectonics, endowing the structural installation with both cultural significance and structural aesthetics. As a result of this research, a new way of interpreting traditional structures has become clear, while a new impetus and direction has been pointed out for structural performance-based design development.

Further research on traditional wooden structures would inevitably point to an ultimate goal: research into the intrinsic material characteristics and behavior of wood. With the development of new timber materials such as composite woods, the characteristics and behavior of wood are constantly changing. We are therefore required to also evolve constantly as the material itself evolves to take advantage of its physical potential through performative design.

参考文献 / References：

[1] 袁烽, 胡永衡. 基于结构性能的建筑设计简史 [J] . 时代建筑, 2014(5): 10–19.

[2] Y.M. Xie, Z.H. Zuo, X. Huang, J.W. Tang, B. Zhao, P. Felicetti: 2011, Architecture and Urban Design through Evolutionary Structural Optimisation Algorithms. International Symposium on Algorithmic Design for Architecture and Urban Design. 2011

[3] 帕纳约蒂斯·米哈拉托斯 著, 闫超 译. 柔度渐变——在设计过程与教学中重新引入结构思考 [J] . 时代建筑, 2014(5): 26–33.

[4] 梁思成. 清式营造则例 [M] . 中国建筑工业出版社, 1981.

[5] 蒋岩, 毛灵涛, 曹晓丽. 古建筑檐椽合理出挑尺寸的结构力学分析[C]. 北京力学会第17届学术年会论文集, 2011: 572–573.

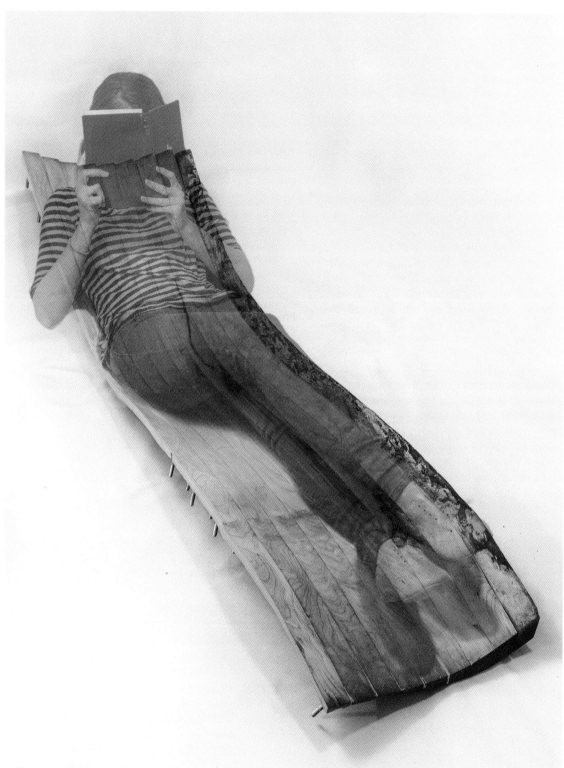

躺椅原型表面。通过合理的利用几何的组合来设计曲面，在生产过程中对材料几乎没有一点浪费。
Prototypical chaise longue surface. Produced with practically zero waste, by strategically designing the surface into the available geometry of the flitch.

Unfolding Topology
Bandsawn Bands
展开的拓扑结构
电锯木条椅的制作

Ryan Luke Johns, Nicholas Foley / Greyshed
瑞安·卢克·约翰斯，尼古拉斯·弗利 / Greyshed设计事务所

建筑设计的第一次数字化转型，部分受到1993年《建筑设计》（AD）杂志的"建筑中的折叠"一文的影响，文中鼓励并倡导连续性、曲线性和可塑性 [1]。数字化和参数化设计工具使得设计达到了前所未有的复杂程度、迭代性和光滑度。虽然建筑师能够在数字世界中塑性地思考和设计，但具体实施却依然相当刻板：一个古板（通常很浪费）的由数控机床定位的几何曲线被强加到直线材料上。尽管数字工具方便了定制化大规模生产，大部分参数化设计加工行为依然局限于标准化工业化构件生产的参数。无论是采用标准化部件叠加装配成"高度信息化"[2]几何造型，还是在部件上通过一系列削减操作来达到效果，最终通用部件（坯料、型材、砖、板材）的可变控制依然不尽如人意。

The first digital turn in architecture, instigated in part by the 1993 *AD* profile "*Folding in Architecture*", encouraged and enabled continuity, curvilinearity, and plasticity [1]. Digital and parametric design tools allowed for an unprecedented level of complexity, iteration, and smoothness. While architects could both think and design plastically in the digital world, any act of materialization generally remained, unfortunately, an act of obstinacy: a stubborn (and often wasteful) CNC-guided imposition of curved geometry onto rectilinear material stock. Despite the capabilities for customized fabrication offered by digital tools, the vast majority of parametric design-fabrication exercises remain confined to the parameters of standard, industrially produced components.

机器人电锯加工
Robotic Bandsawing

为了实现类似于工业化前期手工艺风格的大规模定制,第一次数字转向后的参数化设计师证明有能力以低廉的价格实现变化无穷的几何造型。但手工艺的"自然"性[3]不仅在于匠人手工作品的变化多样,同时还在于加工材料的多样性。匠人并不是在独立的概念空间中设计几何造型然后盲目地应用于材料上,而是将设计深入到可用的材料的本质中去。如要将这种设计手法运用到数字设计中去,我们可以通过算法程序以及数字和物理设计媒介之间的双向转化,将概念上的可塑性运用到已获得的材料条件上。例如,我们并不将一棵弯曲的树转化成一定尺寸的木材,然后再加工成数字化设计的曲线形状,而是直接以树为设计出发点,进入数字化建造。这种制作顺序上的跳跃确保了更加高效可持续地利用材料,同时保留了"天然几何"[4]的美感。

该项目展示了在天然木板的限制条件下,对自由形状表面的设计和制作技术。通过扫描自然弯曲的树木截面,根据其材料形状进行设计,我们能够将一定量紧实的木材展开成一个连续的、拓扑的、建筑尺度的表面。

Whether by creating "highly informed" [2] geometry from an additive assemblage of standardized parts, or from a series of subtractive operations upon them, the variable control provided by the generic component (blanks, bars, bricks and sheet stock) is commonplace.

In attempting to deliver a kind of mass customization reminiscent of pre-industrial craftsmanship, parametric designers following the first digital turn proved the potential for infinite, and cheap, geometric variation. However, the "'natural' uniqueness of craftsmanship" [3] does not arise solely from variations in the hand of the craftsman, but also from variations in the material which is being crafted. Instead of designing geometry in a separate conceptual space, and transferring it blindly into materiality, the craftsman designs it according to the material that is available. In taking this approach to digital design, we can apply conceptual plasticity on to found material conditions through algorithmic procedures and the bidirectional transfer of digital/physical design mediums. Rather than transferring material, for example, from a curved tree into dimensional lumber which is then re-machined

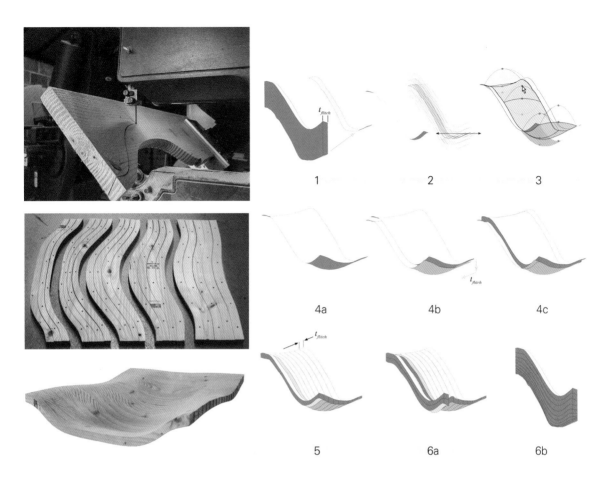

在前期试验中运用固定式带锯机器人对木材进行加工，从规格木条中切割出来的嵌套的木条被组装成夹叠的椅子表面。（左图）
嵌套过程（右图）：1. 组合的木条经过三维扫描和树皮表面建模。2. 树皮表面可以根据所要的几何形体形成的边缘来自由定位。3. 所设计的自由表面从各子集的无限可能性中分离。4. 生成的表皮呈副本状。这一副本在水平向经由木料的厚度进行转译，并向下贴合，由此在两片曲面中创造出一个新的实体（体量等同于操作前的木料）5. 整个体量被横向切分成各种宽度的木条组合。6. 所得到的木条被重新嵌入原始的表皮中，它们将被机器人切割加工。
Through robotic operation, the nested bands were cut from dimensional lumber and assembled laminated Chair surface. (left)
Nesting Process (right): 1. Flitch is 3D scanned and bark surfaces are modelled. 2. Bark surfaces are positioned freely to define the edge conditions of the desired geometry. 3. The designed freeform surface is isolated from an infinite subset of possibilities. 4. The resultant surface is copied. This copy is translated horizontally by the thickness of the flitch, and downwards such that a solid volume is created between the two surfaces (equal in volume to that of the original flitch). 5. That volume is divided horizontally into bands with width flitch. 6. The resultant bands are nested back into the original flitch, to be cut by the robot.

这种工艺采用机器人电锯加工，切割一系列弯曲的木条，然后将木条旋转、层压，制成双曲数字定义的几何体。电锯刀片切口很薄，使得加工表面的嵌套紧密。与计算机数控成形铣削相比，它通过可用材料和设计的几何体之间的紧密联系实现几乎零浪费。非标准材料几何体、3D扫描、参数算法、设计者输入和机器人控制之间的协调加强了基于特征的材料直觉在设计中的作用。

作为少有的不对木材进行削减而是进行变形的木工技术，这种技术的立场显得十分有趣。这种工艺通过参数化生成设计对材料进行重新配置，成为一系列设计师的意愿与材料意愿的折中[5]。

into curvilinear digitally designed geometry, we take the tree as the starting point for design and move directly to digital fabrication. This leap in the production sequence enables more sustainable material efficiency while simultaneously conferring the natural aesthetic advantage of "beauty's found geometries" [4].

This project demonstrates a technique for designing and fabricating freeform surfaces within the constraints of live-edged wood flitches. By scanning a naturally curved tree section and designing within its material morphospace, we are able to unfold a compact amount of wood into a continuous, topological and architecturally scaled surface.

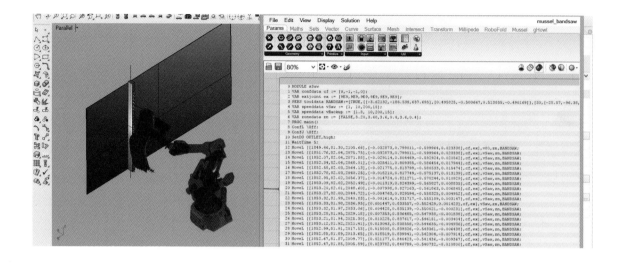

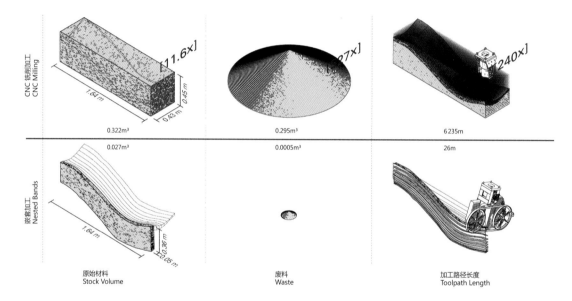

原始材料
Stock Volume

废料
Waste

加工路径长度
Toolpath Length

机器人的加工代码是由"Mussel"生成，它是一个Grasshopper的开源插件。（上图）
实质效率的提高：数控铣削 vs. 样品表面的组合嵌套。（中图）
立式带锯机被重新配置进机器人和感应器中。（下图）
Robot code is generated using "Mussel", an Open Source plugin for Grasshopper. (Above)
Substantial Efficiency Gains: CNC milling vs. Flitch Nesting for a sample surface. (Middle)
Upright bandsaw reconfigured into robot end effector. (Below)

The process utilizes a robotically operated bandsaw to cut a series of curved strips which, when rotated and laminated, can approximate doubly-curved and digitally defined geometry. The thin kerf of the bandsaw blade allows for a tight nesting of finished surfaces, which, through a close relationship between available material and designed geometry, affords practically zero waste when compared to CNC contour milling. This coordination of non-standard material geometries, 3D scanning, parametric algorithms, designer input and robotic control serves to bolster the role of feature-based material intuition in design.

The technique holds an interesting position as one of few woodworking techniques which are explicitly not subtractive, but transformative. By parametrically generating the design as a reconfiguration of the workpiece, the process unfolds as a series of compromises between the will of the designer and that of the material [5].

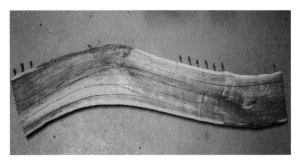

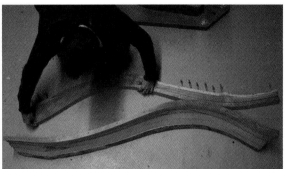

嵌套木条的细节和组装
Details and Assembly of Nested Bands

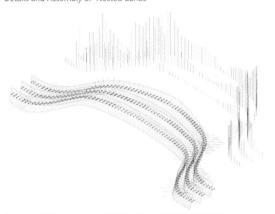

切割速度被进行了优化，在高速旋转时曲率降低以避免断片的产生。
Cut speed is optimized so that movement in areas of higher twist and curvature are slower (to avoid breaking blades).

参考文献 / References :

[1] Carpo, Mario 2004 "Ten Years of Folding." In: Lynn, G (ed) Folding in Architecture: Revised Edition, Wiley.

[2] Bonwetsch, T, Gramazio, F, and Kohler, M 2006 "The Informed Wall: Applying Additive Digital Fabrication Techniques on Architecture", Proceedings of the 25th Annual Conference of the Association for Computer-Aided Design in Architecture, Louisville, Kentucky. pp 489-495.

[3] Gramazio, F and Kohler, M 2008, Digital Materiality in Architecture, Lars Müller Publishers, Baden. pp 10.

[4] Enns, J. 2010 "Beauty's Found Geometries" M. Arch Thesis, Princeton University. Advisor: Prof. Axel Kilian

[5] For more information, see: Johns, R and Foley, N 2014 "Bandsawn Bands: Feature-Based Design and Fabrication of Nested Freeform Surfaces in Wood." In: McGee, W and Ponce de Leon, M (eds) Robotic Fabrication in Architecture, Art and Design. Springer. pp 17-32.

PROJECTS OVERVIEW
项目概况

Studio Roland Snooks & Kokkugia
罗兰德·斯怒克斯工作室 & Kokkugia事务所

RPI & RoboFold
美国伦斯勒理工学院 & RoboFold公司

Gramazio Kohler Research, ETH Zurich & SEC Future Cities Laboratory (FCL)
苏黎世联邦理工学院"格拉马齐奥与科勒"研究中心 & 未来城市实验室

Robothouse, Southern California Institute of Architecture (SCI-Arc)
美国南加州建筑学院机器人实验室

Taubman College, University of Michigan
美国密歇根大学Taubman建筑与城规学院

Hypebody, TUDelft
荷兰戴尔夫特大学Hypebody研究小组

School of Architecture, Tsinghua University
清华大学建筑学院

DigitalFUTURE Shanghai Summer Workshop, Tongji University
同济大学上海"数字未来"暑期工作营

Digital Design Research Center (DDRC), CAUP, Tongji University
同济大学建筑与城市规划学院数字设计研究中心

2015年，复合翼 & 黄铜群在第十届上海双年展城市馆"人机未来"展上展出
Composite Wing & Brass Swarm was exhibited in the "Robotic Future" exhibition of 10th Shanghai Biennale, 2015

Composite Wing & Brass Swarm
复合翼与黄铜群

Roland Snooks / Studio Roland Snooks & Kokkugia
罗兰德·斯怒克斯 / 罗兰德·斯怒克斯工作室 & Kokkugia事务所

建筑空间的可能性是由设计工具来定义和限定的。我们在算法设计上的兴趣点在于其对于重新开发设计工具的能力和重塑设计过程的限定。在近期的实验中，我们通过对工业机器人的使用将算法设计应用到了机器的生产中。这种兴趣是基于工业机器人对于设计末端执行器的开放性，无疑极大程度上推动了建造过程的创造性开发。

我们对算法设计和机器人建造的研究是密切相关的。最初对于机器人的关注是起因于探究实现由算法生成的复杂形体的可能，但现在我们更关注于机器人加工过程和算法行为之间的反馈。通过多代理算法将机器人行为编码，实现了在实际建造环境下得到极多建筑生成的结果可能性。这种算法与机器人行为的互动为建筑实验打开了一片新的天地，并重新定义了建筑师的工具。

The space of architectural possibility is defined and constrained by the tools of design. One aspect of our interest in algorithmic design is the capacity to reinvent design tools and consequently reframe the constraints of our processes. Our more recent experimentation with industrial robotics is a further expansion of the design of tools from the algorithm to the machines of production. This interest is predicated on the openness of industrial robotics - the capacity to design end-effectors and invent fabrication processes.

Our work with algorithmic design and robotics are closely interrelated. While our initial concern for robotics was an interest in fabricating complex algorithmically generated forms, our work is now focused on the feedback between robotic processes and algorithmic behaviors. The capacity to encode robotic behavior into our multi-agent algorithms enables a highly volatile approach to generative architecture while being conditioned by the realities of construction. This interaction of computation and robotics is opening up a new space of architectural experimentation and a reframing the tools of the architect.

复合翼
Composite Wing

设计：罗兰德·斯怒克斯工作室
客户：RMIT设计中心
设计团队：罗兰德·斯怒克斯（负责人），加姆·纽汉姆，德鲁·布斯迈，阿姆瑞·托马斯，李培舍
结构工程师：Bollinger+Grohmann 工程公司
支持：皇家墨尔本理工大学，d_Lab机器人建造及设计工作营
Design: Studio Roland Snooks
Client: RMIT Design Hub / Fleur Watson + Kate Rhodes
Design Team: Roland Snooks (Director), Cam Newnham, Drew Busmire, Amaury Thomas, Pei She Lee.
Structural Engineering: Bollinger+Grohmann Engineers
Supported by: RMIT University, d_Lab.Architectural_Robotics, and Architecture and Design Workshops

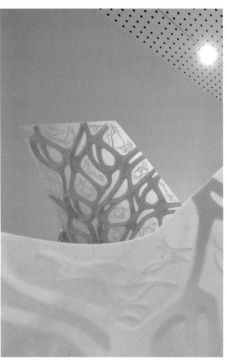

复合翼是罗兰德·斯怒克斯工作室完成的一个装置，该装置结合了算法设计和机器人建造，实现将错综复杂的脉网嵌入在单薄的复合表面内。这个装置探讨了一种建筑原型，探索如何将表面、结构及装饰压缩为单一不能复归的形态。

该项目为罗兰德·斯怒克斯工作室和Kokkugia研究实验室进行的算法、材料及机器人设计研究的延伸。表面主要采用玻璃纤维材料，装饰/结构镶嵌物部分由机器人进行建造。机器人建造技术使得实现该项目复杂的图案及几何结构成为可能，包括微型表面接合的挤制。根据表面接合的位置及形式探讨结构强度，将脉络作为嵌入在表面内的结构柱，这种策略使得表面在保持仅数毫米厚度的同时，还能具有较大距离的跨度及悬臂。

表面内错综复杂的图案通过多代理算法进行设计（多代理算法是一种基于集群智能的自组织逻辑）。生成方式则结合了结构、形式及装饰设计意图，试图得到一种无法预期的形式结果。这个项目试图脱离构造元件的分离式节点形式，而是以一种系统式的压缩方式表达。

Composite Wing is an installation designed by Studio Roland Snooks that combines algorithmic design and robotic fabrication. The installation comprises an intricate network of veins embedded within a thin composite surface. It is an architectural prototype that explores how surface, structure and ornament can be compressed into a single irreducible form.

The project is an expansion of the algorithmic, material and robotic design research of Studio Roland Snooks' research lab, Kokkugia. The surface is fabricated primarily from fiberglass with the ornamental/structural inlay produced robotically. The complex patterns and geometry of the project are made possible through custom robotic fabrication techniques including the extrusion of the fine-scale surface articulation. The project gains its structural strength through the location and pattern of this surface articulation, with the veins operating as structural beams embedded within the surface. This strategy enables the surface to remain only a few millimeters thick while spanning and cantilevering considerable distances.

The intricate pattern within the surface is designed through a multi-agent algorithm that draws from the self-organizing logic of swarm intelligence. This generative approach negotiates between structural, formal and ornamental design intentions in creating an emergent surface condition. The project resists the separate articulation of tectonic elements, instead these are compressed into a synthetic whole.

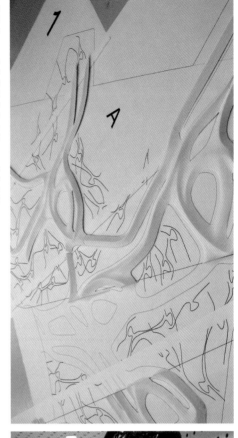

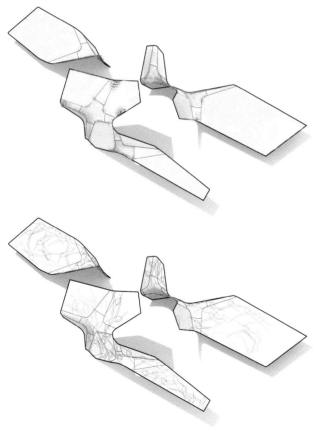

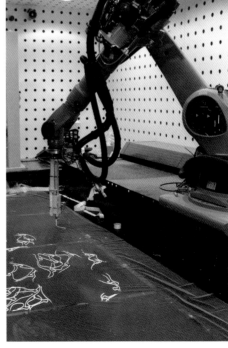

复合翼的结构分析
Structural Analysis, Composite Wing

复合翼的制作和安装过程
Fabrication and Installation Process, Composite Wing

黄铜群
Brass Swarm

设计：Kokkugia事务所 | 罗兰德·斯怒克斯
设计团队：罗兰德·斯怒克斯，加姆·纽汉姆，德鲁·布斯迈
支持：皇家墨尔本理工大学，d_Lab机器人建造及设计工作营
Design: Kokkugia | Roland Snooks
Design Team: Roland Snooks, Cam Newnham, Drew Busmire
Supported by: RMIT University, d_ Lab.Architectural_Robotics, and Architecture + Design Workshops

黄铜群是基于自组织算法设计及机器人建造互动的设计过程而研发的一种试验模型。本项目旨在探索空间自组织、涌现建构，及机器人行为与算法行为之间的关系。

Kokkugia事务所开发了拓扑表皮空间的自组织多代理算法策略。这种多样性的智能化策略，将大量的代理元进行自组织的云计算，形成相互的连续表面及复杂的空间区域，同时生成本错综复杂的最终建构形式。每个代理都有一个主体可与周边代理主体互动和联系，这些代理主体的互动形成错综复杂的装饰性及结构网络。

两个Kuka Agilus机器人的充分互动形成了这个项目的原型，将黄铜杆件弯曲作为代理元件。机器人包含一套精确的物理机械约束部件，这些部件限制潜在的弯曲角度和方向。根据生成算法规则对机器人的这种行为进行编码，以确保代理主体将高度变化性的生成设计行为与更多制造考虑因素联系起来。整个设计方法将建造和设计同时进行，而无需在后期进行按照建造标准将复杂的几何体合理化的步骤。

这些多代理算法过程是Kokkugia研发的行为生成策略的一部分。这种策略通过将建筑设计决定编码，促发计算代理元间的互动和一种自组织的设计意图及一种涌现的建筑形式。

Brass Swarm is an experimental prototype developed through self-organizational algorithmic design processes and robotic fabrication. The project explores spatial self-organization, emergent tectonics and the relationship between robotic and algorithmic behavior.

Kokkugia has developed a multi-agent algorithmic strategy for spatial self-organization, from which topological surfaces emerge. This manifold swarm strategy self-organizes clouds of agents into coherent, continuous surfaces, and complex spatial division. These spatial agents simultaneously generate the intricate tectonics of the project. Each agent has a body that is capable of interacting and connecting to the bodies of the surrounding agents. The interaction of these agentBodies generates intricate ornamental and structural networks.

The prototype is robotically fabricated through the interaction of two Kuka Agilus robots, which bend the brass rods comprising the agentBodies. The robots have a precise set of physical and mechanical constraints that limit the possible angles and direction of the bends. This behavior of the robots has been encoded within the generative algorithm to ensure that the agentBodies negotiate between highly volatile generative design behaviors and more pragmatic concerns of fabrication. This process compresses fabrication and design imperatives into a single process, removing the need to post-rationalize complex geometry with regard to fabrication criteria.

These multi-agent algorithmic processes are part of the Behavioral Formation strategy developed by Kokkugia. This strategy encodes architectural design decisions within computational agents that interact, giving rise to a self-organised design intention and an emergent architecture.

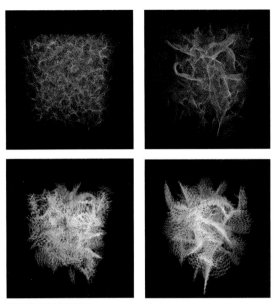

黄铜群的自组织算法设计
Self-organisational Algorithmic Design, Brass Swarm

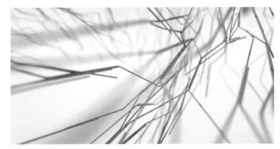

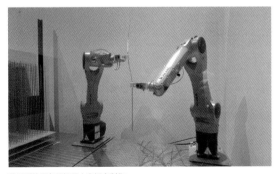

黄铜群的铜条用机器人弯折来制作
Robotic Bending, Brass Swarm

项目团队

RPI团队：萨哈·米汉多斯特，郭寰宇，杰西卡·科利尔，伊丽莎白·萨马提诺，马修·沃杰尔
RoboFold团队：格里高利·艾比思，艾玛·艾比思，弗洛伦特·米歇尔，杰格·杜德利

* 本项目是安德鲁·桑德斯与RoboFold工资的合作项目，得到了2013年伦斯勒理工学院"罗伯特 S. 布朗研究基金"的支持

Project Team

RPI Team: Sahar Mihandoust, Guo Huanyu, Jessica Collier, Elizabeth Sammartino, Matthew Vogel
RoboFold Team: Gregory Epps, Ema Epps, Florent Michel, Jeg Dudley

* Andrew Saunders in collaboration with RoboFold Ltd.
Sponsored by the Rensselaer 2013 The Robert S. Brown '52 Fellows Program

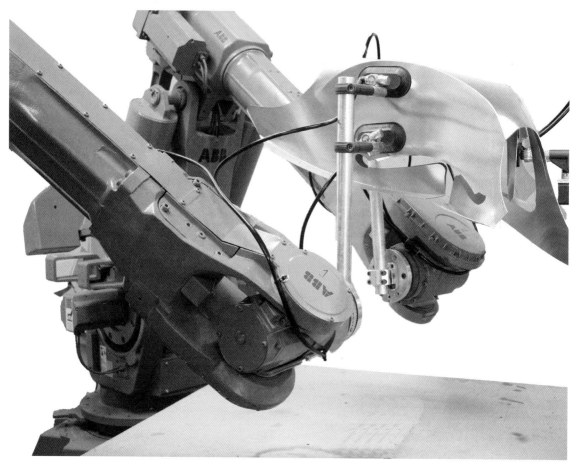

机器人金属弯折
Robotic Banding

Robotic Lattice Smock
机器人点格褶裥

Andrew Saunders (RPI) & Gregory Epps (RoboFold)
安德鲁·桑德斯（美国伦斯勒理工学院）& 格里高利·艾比思（RoboFold公司）

19世纪德国建筑师及历史学家戈特弗里德·森佩尔声称建筑是一系列构件连接围护出来的空间。机器人点格褶裥绣（RLS）作为对于这个概念的现代演绎形式，通过计算模拟及机器人建造流程将点格褶裥绣对可堆叠材料建构的影响转移至折叠弯曲的平面金属板。

该流程的第一步是对毛毡制品进行缩褶处理。由于毛毡制品具有相对的刚性，在对毛毡织物进行缩褶时会形成独特的折纹及摺痕。一旦这些折纹集中出现在毛毡制品上，就进行缝合，并将毡制品恢复至扁平板上，进而在原有的网格上形成新的弧形折叠图案。

将重纸板切割成新的扁平图案。展开这些模型，并在扁平图案上的标准线处进行精确计算，模拟折叠流程。采用Kangaroo的Live Physic工具（丹尼尔·派克研发的Grasshopper插件）在数字化背景中精确塑造并模拟，将扁平刨机材料转换成最终的折叠形式流程。

在本研究中，运用Kangaroo软件的数字模拟来精确设计多个机械臂的移动，进而实现折叠。运用Godzilla（一款由格里高利·艾比思研发的Grasshopper插件）操作机械臂模拟折叠流程。这样，就可优化箭状褶裥最终形成的扁平图案，并调整其大小以便设置机器人夹具。

Gottfried Semper, the nineteenth-century German architect and historian asserts that textiles are the origins of buildings. As a contemporary projection of this framework, Robotic Lattice Smock (RLS) transposes the affects of pliable material tectonics of lattice smocking to folding and bending of planar sheet metal through computational simulation and robotic fabrication processes.

The process begins with smocking felt. Felt is chosen for its relative rigidity so that distinct fold lines and creases emerge when the fabric is smocked. Once these lines are traced on the felt in its gathered state, the stitches are released and the felt returns to a flat sheet, revealing new curved fold patterns on the original grid.

The new flat patterns are used to cut heavy weight paper templates. These models are unfolded and the locations of ruling lines on the flat pattern are used to construct a precise computational simulation of the folding process. To accurately model and simulate the transition from flat planer material to final folded form in the digital environment, the Live Physic engine Kangaroo (Grasshopper plug-in developed by Daniel Piker) is used.

The Kangaroo digital simulation is used to choreograph the exact movement of multiple robotic arms to achieve the fold. Once

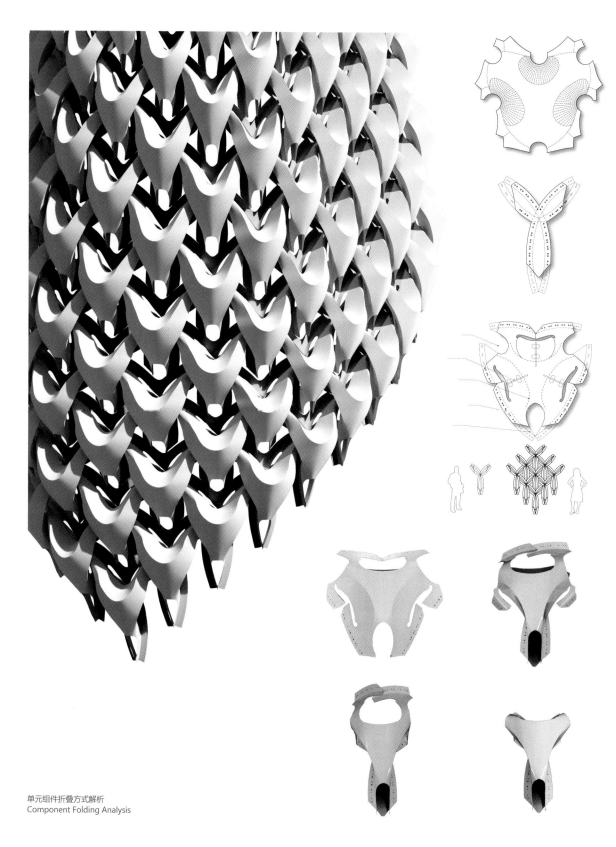

单元组件折叠方式解析
Component Folding Analysis

the folding process is simulated with robotic arm placement with Godzilla (Grasshopper plug-in developed by Gregory Epps), the final flat pattern for the arrow smock is modified and resized for positioning of robotic grippers.

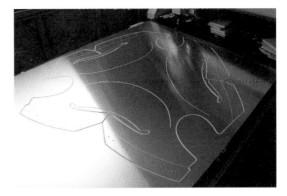

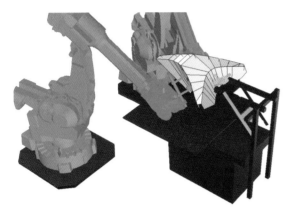

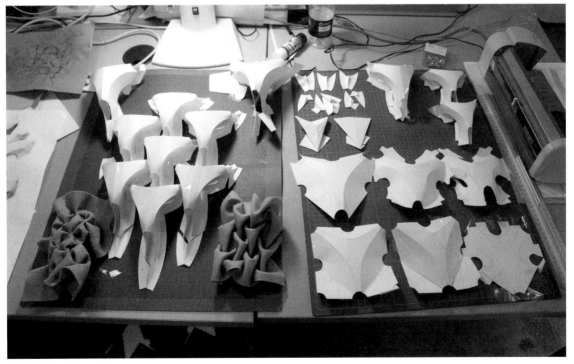

模型建造
Mock-up Fabrication

机器人建造高层建筑设计2

合作者：
迈克尔·布迪格（项目主持），赛伦·埃尔詹，诺曼·哈克，史莱克·兰格博格博士，威利·劳尔，依米尔科·利姆，杰森·利姆，拉斐尔·派托维克

学生：
塞巴斯蒂安·恩斯特，帕斯卡·根哈特，帕特里克·戈登那，塞尔维斯·卡拉马，斯文·洛克霍夫，斯尔文·斯托巴克，迈克尔·斯顿兹，弗洛伦斯·托尼，马丁·特沙兹，阿尔瓦托·瓦卡斯，法比恩·沃德博格，托比斯·沃什尔格，皮托斯·阿梅鲁斯-林德斯托，陈彭宏，冯凯辉，何宇航，大卫·珍妮，侃利金，李平凡，简-马克·斯坦利曼，安德鲁·王

Design of Robotic Fabricated High Rises 2

Collaborators:
Michael Budig (project lead), Selen Ercan Norman Hack, Dr. Silke Langenberg, Willi Lauer, Emiko Lim, Jason Lim, Raffael Petrovic

Students:
Sebastian Ernst, Pascal Genhart, Patrick Goldener, Sylvius Kramer, Sven Rickhoff, Silvan Strohbach, Michael Stünzi, Florence Thonney, Martin Tessarz, Alvaro Valcarce, Fabienne Waldburger, Tobias Wullschleger, Petrus Aejmelaeus-Lindström, Pun Hon Chiang, Kai Qui Foong, Yuhang He, David Jenny, Lijing Kan, Ping Fuan Lee, Jean-Marc Stadelmann, Andre Wong

机器人建造高层建筑设计2
Robotic Fabricated High Rises 2 © Gramazio Kohler Research, ETH Zurich

Design of Robotic Fabricated High Rises
机器人建造高层建筑设计

Gramazio Kohler Research, ETH Zurich & SEC Future Cities Laboratory (FCL)
苏黎世联邦理工学院 "格拉马齐奥与科勒" 研究中心 & 未来城市实验室

新加坡国立大学和苏黎世联邦理工学院的联合教学中心——"未来城市实验室"（FCL），是由苏黎世联邦理工学院建筑数字化建造研究中心的法比奥·格拉马齐奥和马蒂亚斯·科勒教授建立，用来研究机器人在建造和设计东南亚高层建筑类型潜力的前沿机构。其中Module II研究团队成立了一个特别实验室，专门研究如何通过将算法设计与机器人技术相结合，脱离目前占主导地位但已过时的建筑形式，同时探索规模经济、序列化及标准化结构在建筑中的呈现。研究第二年，研究团队展开"机器人建造高层建筑设计2"的设计项目，2013年，研究团队与新加坡国立大学建筑学院合作组织展开本项目研究。第一年的"机器人建造高层建筑设计1"集中探索机器人在1：50模型建造过程中的材料定义及建造过程；第二年，研究团队基于先前的经验着重探索建筑类型的演化发展。

In the context of the Future Cities Laboratory (FCL) at the Singapore-ETH Centre (SEC), the Professorship for Architecture and Digital Fabrication of Fabio Gramazio and Matthias Kohler has built up a unique laboratory to research the potentials of robotic processes for the design and construction of high rise typologies in Southeast Asia. The research team of Module II is representing the FCL and has built a unique laboratory to investigate how computational design, coupled with robotic technologies has the potential to liberate this interesting building type from its still dominant, but culturally and technologically obsolete subjectification to the economies of scale, of serialization, and standardization. Design of Robotic Fabricated High Rises 2 is being conducted for the second consecutive year. In 2013, it was organized as collaboration with the Faculty of Architecture of the National University of Singapore (NUS). While the first year Robotic Fabricated High Rises 1 focused on experimentation with the robotic facilities in order to define material and fabrication processes that are suitable for 1:50 scaled model building, the second year builds upon previous experiences and is more geared towards an evolution of architectural typologies.

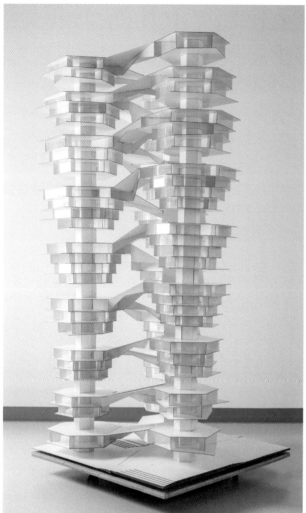
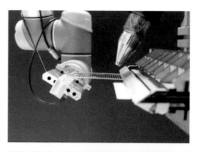
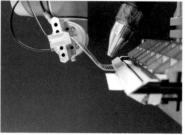
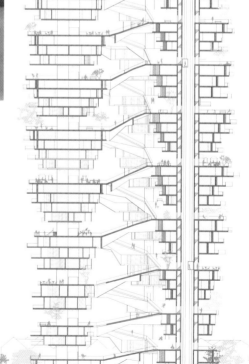

"垂直大道"：该项目提出了一个可以提供公共空间和垂直花园的螺旋式上升的循环系统。整个建筑由三栋集中在核心筒区域的高层公寓组团组成。一个交错开放的公共空间系统由一个连续的螺旋上升的斜坡连接，可作为一个公共的会面和公园空间。

Vertical Avenue: This project proposes a spiraling circulation system that provides public programmes and parks vertically throughout the tower. The building consists of three high-rise towers with apartment clusters around the central cores. A system of staggered open public spaces is connected by a continuously upward-spiraling ramp, which serves as a public meeting space and park.

© All images by Gramazio Kohler Research, ETH Zurich

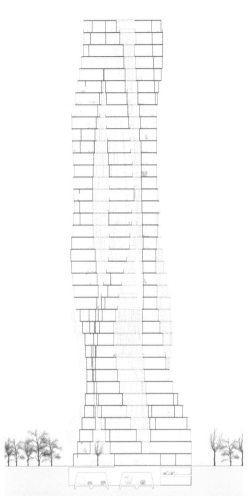
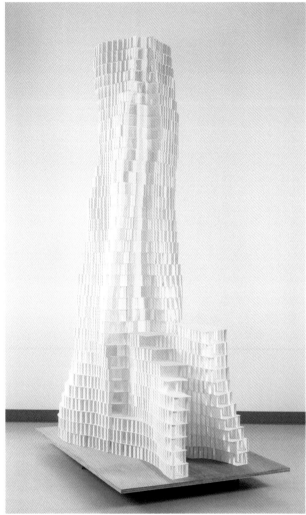

"连续帧"：该塔楼意图通过将算法设计运用到极致，在几何元素之外创建一种拥有众多独特室内空间的体验；计划将塔楼作为一种连接手法，通过分开和聚合来形成一个起伏的整体形状，从而连接了新加坡的公路和两个相邻的公园。

Sequential Frames: The design intention of this tower is to create a multitude of unique interior spatial experiences out of simple geometrical elements, by deploying the full power of computation. The towers are planned as linked strands that branch and merge into an undulating overall shape, bridging a Singaporean highway and connecting two adjacent parks.

© All images by Gramazio Kohler Research, ETH Zurich

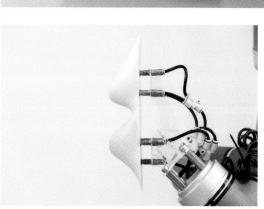

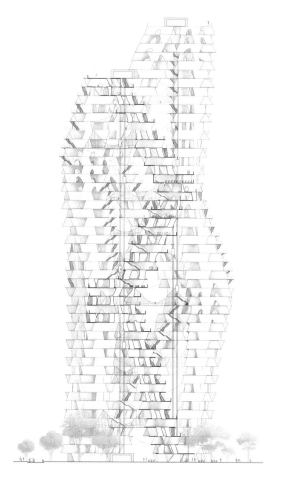

网格塔：该项目提出了将一种多孔网状的细长塔楼来作为一个巨大的公寓集合体的选择性方案，社区邻里直接与一个新加坡的保护区毗邻。整体形态的建筑占地被最小化，从而在地面上为公园腾出了空间，并与周围环境的历史结构形成了互补。

Mesh Towers: This project proposes a porous mesh of slender tower strands as an alternative to a massive condominium project, planned adjacent to a conservation area of Singapore. The footprints of the overall shape are minimized, keeping the ground level open for a park and complementing the historical structure of its surroundings.

© All images by Gramazio Kohler Research, ETH Zurich

在2013年的"网格模型"项目中，一个移动的机器人制作了一个有一定高度的原型
Mesh Mould, 2013: Tall prototype extruded with a mobile robot. © Gramazio Kohler Research, ETH Zurich

Mesh Mould
网格模型

Gramazio Kohler Research, ETH Zurich & SEC Future Cities Laboratory (FCL)
苏黎世联邦理工学院"格拉马齐奥与科勒"研究中心 & 未来城市实验室

在"网格模型"项目中,加固及模板制作这两个独立要求在现场机器人制造过程中得以合二为一。现场直接挤压模板在简化流程的同时,可应对更复杂的几何结构。由于材料用量被降为最低,此种方式最大创造了能效。在当前情况下,机器人挤压工艺可处理各类热塑性塑料材料。从熔融层积到空间挤出的打印概念转变有重要意义:前者为通用型,多数用于形式的表现;而数字控制的空间挤出则可针对不同的建筑结构,并可同时减少生产时间及产品重量。Sika Technology AG作为合作伙伴及水泥材料专家,协助了项目(美国临时专利申请号:61/873,467)的完成。

In Mesh Mould, two separate requirements – reinforcement and formwork – are folded into one single in situ robotic fabrication process. The direct extrusion of the in situ formwork allows for a greater geometric complexity while simplifying the process itself. Since the amount of the required material is reduced to a minimum, such an approach holds a high potential for resource efficiency. At our current conditions, the robotic extrusion technique can process different kinds of thermoplastic materials. The conceptual change from layer-based deposition to spatial extrusion has noteworthy implications. Whereas the former remains generic, mostly for the representation of form, the digitally controlled spatial extrusion becomes specific to the architectural construction and allows for a significant and simultaneous reduction of both weight and production time. The project (US Provisional Patent Application No. 61/873,467) is conducted in close collaboration with Sika Technology AG as an industry partner and expert in cement-based materials.

网格模型
Mesh Mould

合作者：诺曼·哈克（项目主持），威利·维克多·劳尔
工业合作伙伴：Sika技术公司
Collaborators: Norman Hack (project lead), Willi Viktor Lauer
Industry Partner: Sika Technology AG

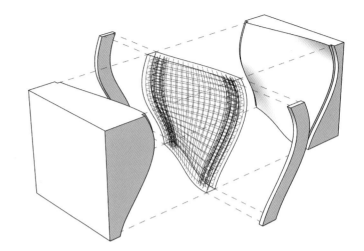

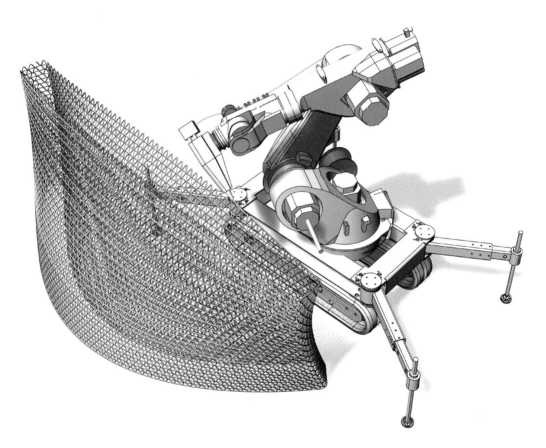

混凝土结构所需费用中超过60%的都来自于模具的加工和不同模块之间的连接，Mesh Mould将这两个系统组合成了一个框架加强系统。这种3D网络结构可以通过机器人打印出来，这种方法不会产生任何垃圾，并且能在不产生额外成本的情况下提供足够的几何复杂性。
Over 60 per cent of the costs of a concrete structure are due to the labour-intensive construction of formwork, and bending and placing of reinforcement accounts for another significant share. Mesh Mould proposes the unification of these two systems into one combined formwork-reinforcement system. The 3D mesh structures are robotically fabricated in an additive, waste-free manner, providing increased geometric complexity without raising the costs. © All images by Gramazio Kohler Research, ETH Zurich

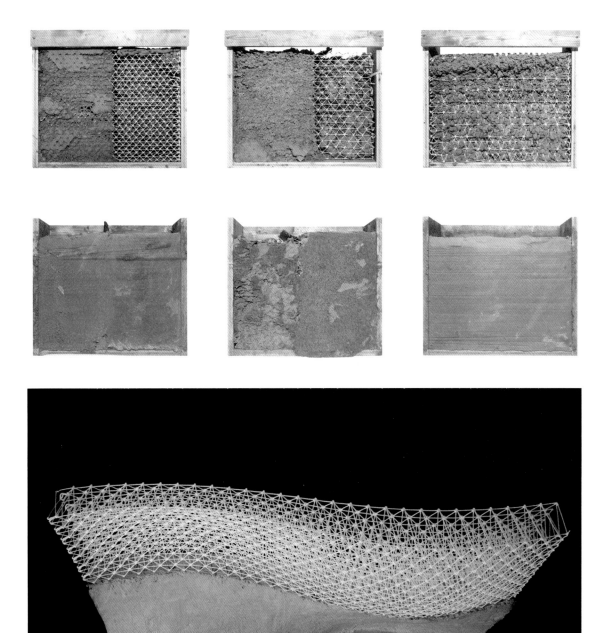

一些大尺寸（80cm×60cm×8cm左右）的ABS打印样品用来测试这个方法的稳定性。
A few large acrylonitrile butadiene styrene (ABS) samples with sizes of approximately 80cm×60cm×8cm were fabricated to test the robustness of the process. © All images by Gramazio Kohler Research, ETH Zurich

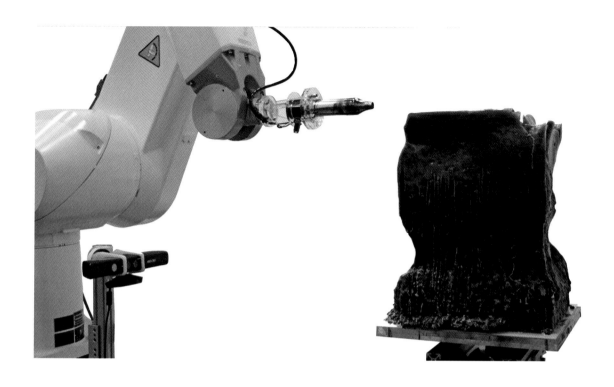

2014年"自动：设计智能化"的课程中的组2和组6的作品
Group 2 & Group 6, Automaton: Designing Intelligence, 2014

Automaton
Designing Intelligence
自动
设计智能化

Robothouse, Southern California Institute of Architecture (SCI-Arc)
美国南加州建筑学院机器人实验室

本项目运用机器人控制平台中的智能代理方法，来探索实时响应系统的内在潜能及问题。所有项目基于Processing开发的控制平台操作，通过一系列分析及决策功能，与指令（此处指代微软Kinect及机器人于现实世界中的位置）进行连接，随后这些指令被转化成为控制机器人运动的数据。本项目旨在探索这些系统的潜能，将设计师的意图拓展到这些极端情况或动态材料互动的场景。这些场景要求采用非线性方法，并将启发式设计与建造行为进行归并。

在本学期南加州大学（SCI-Arc）机器人实验室的教学中，最常见的是利用软件平台（如Maya或Grasshopper）模拟完整的运动路径，然后将该路径输出至机器人终端。该过程确保设计师可针对最终预期结果研究并预测机器人的全部运动序列。该过程的拟模型利用Staubli机器人的及时反馈功能，针对紧急受限结果探索由下而上/非线性设计方法。与通过设定机器人运动路径来满足预期设计目标不同的是，学生们基于对智能代理的研究，利用上述过程设计出机器人面对不断变化的环境做出的响应行为。根据设计的特定目标来建造的这一行为越发

The project looks at utilizing intelligent agency within the robot control platform to explore the potentials and issues inherent in real time responsive systems. All projects leveraged a control platform developed in Processing to link precepts (in this case a Microsoft Kinect, and the robot's real world position) through a series of analytical and decision making functions, which are then translated into robot motion control data. The intention of the project is to explore the potentials of these systems to extend the designer's intentions into situations which deal with extreme contexts or dynamic materials – situations which require non-linear methods and the conflation of design heuristics with the act of fabrication.

The most common model utilized within the Robot House at SCI-Arc prior to this seminar, was to simulate a full motion path on a software platform like Maya or Grasshopper and then export that path to the robot's controller. This model allowed the designer to study and predict the robot's entire motion sequence towards an intended final outcome. The model proposed by this course

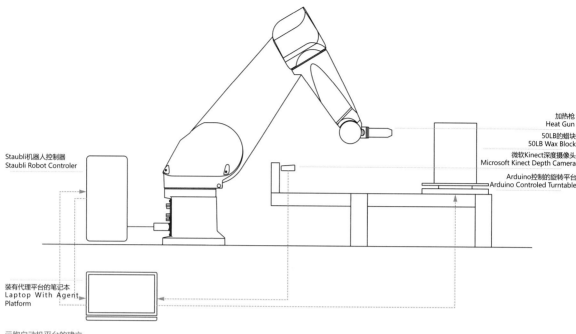

元胞自动机平台的建立
Automaton Platform Setup

罕见，设计师开始倾向于将设计意图及逻辑理解融入到建造装置中。设计师的决策被扩展到装置中，通过非人类能力解释材料及环境并与之响应。项目同时对设计过程中的模拟/表现及建造过程进行分解，为建筑师和设计师提供了其他运行模式。

机器人运动激发平台是由凯西·莱姆根据Processing集成开发环境开发出来的。除了凯西·莱姆开发的Processing及功能中的数据库，平台中用了下列第三方库：toxicLibs用于矢量数学及网格生成；simpleOpenNI用于与Kinect的连接；Obsessive Camera Direction用于Processing开发界面中的模拟摄像机。同时Arduino被用来控制外部硬件。由杰克·纽斯曼开发的Staubli's Val3脚本语言中的程序在机器人控制器中运行，在通过传输控制协议连接听取更新后的机器人运动路径的同时，以数据流的形式持续输出机器人当前的位置及方向。

这一过程中开发的所有代理程序均为"响应式代理"，与真实世界保持同步[1]。代理软件不具有任何学习或适应能力。结果中的非线性特征由整套系统得出，包括熔蜡、传感设备、代理软件及蜡的机器人操控。代理软件不间断听取机器人的位置信息，在机器人达到原指定位置后，对蜡进行扫描，并在一系列行为（分为三部分：靠近、应用及返回原位置）完成后，产生一套新的运动路径。系统在激活后，无需用户输入即可自动运行。

学生们使用了多种方法分析扫描蜡，根据扫描结果生成工具路

leverages the real time streaming capability of the Staubli robots to explore bottom up/ non-linear design methods towards emergent formal outcomes. Instead of designing an intended outcome and then developing robot motion paths to achieve that goal, students design the behavior of the robot's response utilizing models based on research into intelligent agents to engage constantly changing contexts. This exercise becomes less about designing a specific object and then fabricating it. Rather, the designer codifies his/her intentions and formal intelligence into the fabrication device, the goal being to extend a designer's decisions into a device which can interpret and engage materials and contexts with non-human capabilities. The project also collapses simulation/representation and fabrication within the design process, offering an alternative mode of operation for architects and designers.

The platform for generating the robot motion was developed by M. Casey Rehm utilizing the Processing IDE. In addition to libraries native to Processing and functions developed by M. Casey Rehm, the following third party libraries were utilized in the platform: toxicLibs for vector math and mesh generation, simpleOpenNI for interfacing with the Kinect, and Obsessive Camera Direction for simulated cameras within the Processing developed interface. Arduino was used to control external hardware. A program in Staubli's Val3 scripting language developed by Jake Newsum runs on the robot's controller, continuously streaming out the robot's

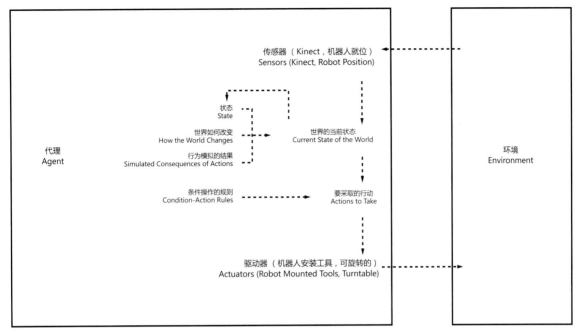

基于条件反射的代理模型
Model Based Reflex Agent - from Russell and Norvig (1995)

径。这些路径分成两类：单一代理及多代理运算。第1组利用单一代理方法，仅仅凭借网格的几何特性即生成运动路径。通过附加的叠加反馈行为，小组希望增强表面动态。扫描的网格集合体及机器人路径的直接联系保证了"安全"行为的产生。这种行为确保机器人不会移动至未经扫描的网格区域，但也将几何路径限定在了三角形网格边缘的对角线中。

其他小组受到克里格·雷诺德 [2] 及约翰·霍兰德 [3] 工作启发，使用多代理系统开发出叠加分析层，用来"读取"被扫描的网格表面。大量相对非智能的代理元穿过网格，随后根据一套研究标准从以往路径中选择最适宜的路径。这一做法使实际工具路径及材料表面之间的关系变得紧张，因此编排及序列问题就变得非常关键。由于这些模型依靠繁杂的计算及不断变化表面中的模拟运动轨迹，因此需开发出方法，处理潜在的不协调问题。第6组学生在每次扫描后及运行多代理模拟的同时中止机器人的运动，直至产生所需的紧急结构层。从而在下次扫描开始前，从单一代理的轨迹中获得运动路径，并清除点数。在高效控制路径及当前几何体的紧密关系时，该流程要求制造过程有2min多的延迟，以完成所需计算。

第4组保留了存储着历史位置信息的云代理元的活性群体。在进行扫描时，当前及此前位置中的代理被推入更新后的网格中。虽然这一行为可缩短计算时间，但网格扫描中的错误将导致代理位置被不可预见地更新。第4组的代理方案不能很好地

current position and orientation while listening for updated robot motion paths via TCP connection.

All agent programs developed in this course are reflex agents that keep track of the world as described by Russell & Norvig (1995) [1]. The agent software does not employ any learning or adaptive capabilities. The non-linear characteristics of the outcome derive from the complete system including melting wax, sensory equipment, agent software and robot manipulation of the wax. The agent software continuously listens for the robot's position. Once the robot has achieved a specified home position, a new scan is taken of the wax and new set motion paths are generated usually following a three part series of 'approach', 'apply', and 'retreat back to the home position'. Once activated the system runs autonomously without user input.

Several methods were utilized by the students to analyze the scanned wax and generate tool paths from the scans. These fell under two categories: single agent and multi agent algorithms. The single agent method utilized, for example, by group 1 relies exclusively on the geometric properties of the mesh to generate the motion path. By pursuing a behavior resulting in additive feedback, the group hoped to produce increased surface dynamics. The direct relationship between the scanned mesh's geometry and the path of the robot resulted in a "safe" behavior

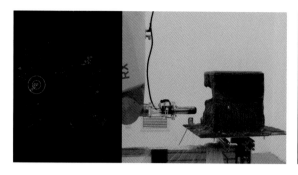

1. 对蜡块进行扫描和网格定位，将离机器人最近的网格点作为加工的起点。
Scan the wax and locate mesh vertex closest to robot origin and set as the initial tool position.

1. 环境交互因子根据网格面的热学效应形成轨迹。
Stigmeric agents generating history based trails run contiously on the last scan's mesh. The agents have an additional behavior with an attraction to recently heated area's of the mesh.

2. 测试该点与相邻点的连线在该点上的法线。
Test close vertex normal against edge connected vertex normals.

2. 在产生新的刀路之前，新的扫描被捕捉，交互因子和交互的历史被投射到新的网格上。
Prior to generating a new set of tool paths, a fresh scan is captured, the agents and their history are projected onto the new mesh.

3. 移动工具至法线角度最大的点，重复这个过程00次，重复过程中不回到前一个点。
Move tool position to vertex with greatest normal angular difference. Repeat 80 times without returning to a previous vertex.

3. 临近位置的交互因子被选中，一系列的同心圆被创建，通过连接各因子的现在与过去的位置，以确定机器人刀路。
The agent with the closest position and who's trails are entirely within the work area is selected. A series of concentric circles is created for the robot's toolpath by linking the agent's current position with positions along its history.

组1：加工过程 & Pseudo代码
Group 1 : Fabrication Process & Pseudo-Code

组2：加工过程 & Pseudo代码
Group 2 : Fabrication Process & Pseudo-Code

处理多个与项目限制相关的问题。作为代理行为的一部分，他们建立了模拟材料温度的计算过程，以吸引代理元运行路径到达最近加热过的区域，从而加速变形过程。此外，他们开发出纤维工具行为，与热风枪产生的固有变形形状形成了互补。

未来对平台的研究将重点关注精细工具的开发，以确保加减法动作同时进行的可能性，同时满足不同尺度的构件连接。目前，当软件计算出现复杂的几何潜力时，材料操作方法却只能限制于非常粗糙的尺度进行操作。未来还将针对行为标准及适应性行为深入探索，从而将代理元的智能性特质拓展至建筑建造过程中。

which would never tell the robot to move into an unscanned area of the mesh. However it also limited the path geometries to the inherent diagonals of triangulated mesh edges.

Other groups developed an additional layer of analysis, utilizing multi-agent systems inspired by the work of Reynolds (1987) [2] and Holland (1995) [3] to "read" the scanned mesh surface. A large number of relatively unintelligent agents were run across the mesh, then a search criterion was used to select appropriate paths from their histories. While this generated more emergent relationships between the actual tool path and the material surface, issues of choreography and sequence became critical. As these models relied on a heavier calculation and a history of simulated motion across a constantly changing surface, methods were developed to deal with potential dissonance. Group 6's program paused the robot motion after each scan while running several generations

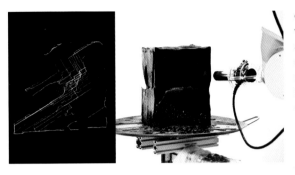

of agent simulations until a desired level of emergent structure was produced. Then a motion path was extracted from a single agent's history and the population was cleared until the next scan. While effective in keeping a tight relationship between paths and current geometry, this process required delays of over 2 minutes in the fabrication process to complete the necessary calculations.

Group 4 maintained a continuously active swarm in which each agent maintained a past history of positions. As new scans were taken, the current and past positions of the agents were snapped onto the updated mesh. While this alleviated issues relative to computation time, errors in mesh scans could result in unexpected updates to agent positions. Group 4's agent program did do several things well in respect to the project constraints. As part of their agent behavior group 4 built in a simulated material temperature calculation to attract the agent paths to areas which had been recently heated in order accelerate deformation. Additionally, they developed a funicular tool behavior which complemented the inherent deformation shape created by the heat gun.

Future research into the platform will focus on the development of more sophisticated tooling to allow both additive and subtractive operations to be used in concert, and to allow for articulations at multiple scales. While the software is beginning to produce sophisticated geometric potentials, the methods of material manipulation currently operate at too crude a scale. Additional explorations will also look at the exploration of performative criteria and adaptive behaviors to extend the agent intelligent towards architectural consequences.

1. 每一步都要对蜡块进行扫描和网格定位。
A clean scan of the wax mesh is taken every sequence.

2. 50个环境交互因子形成了网格的吸引力，角的约束及在新的网格上形成了100次迭代的寿命轨迹。
50 generations of stygmergic agents with mesh attraction, angular constraint, and a trail lifespan of 100 iterations are run on the new mesh.

3. 机器人的轨迹根据交互因子在机器人安全工作范围内的最长轨迹而设定。
The robot's tool path is plotted along the longest agent trail which remains within the robo's safe work area.

组6：加工过程 & Pseudo代码
Group 6 : Fabrication Process & Pseudo-Code

参考文献 / References：

[1] Norvig, Peter, and Stuart J. Russell. Artificial Intelligence: A Modern Approach. New Jersey: Prentice Hall (1995).

[2] Reynolds, Craig. "Flocks, Herds, and Schools: A Distributed Behavioral Model." Computer Graphic, Vol. 21 (1987).

[3] Holland, John. Hidden Order: How Adaptation Builds Complexity. Basic Books (1995).

项目团队

指导老师：彼得·泰斯特，克瑞姆·巴特莱纳，迈克尔·杰克·纽斯曼
学生：戴森，魏娉婷

Project Team

Tutors: Peter Testa, Curime Batliner, Michael Jake Newsum
Students: Dai Sen, Wei Pingting

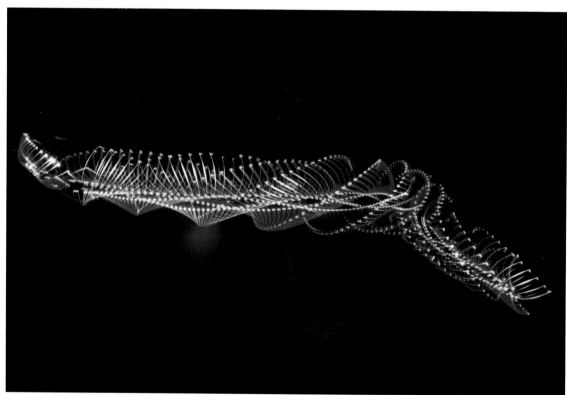

光物体
Light Object

Gesture translation & Light Object
手势转译与光物体

Robothouse, Southern California Institute of Architecture (SCI-Arc)
美国南加州建筑学院机器人实验室

与按顺序衔接不同的工作流程不同的是，本设计在于探索整合3D到2D和2D到3D的新的工作流程。这个目标摈弃了传统的将电脑视为2D的输入、输出装置，而将它看作是连接两个物理世界的桥梁。新的设计模式打破原有的层级关系，创造了平行和整合的设计。

机械臂的意义在当它们从工厂生产线到SCI-Arc的机器人实验室时已经发生了改变。我们跳出将机械臂视为一种遵循代码重复劳动的机械装置的传统，机械臂不再是顺从的"奴隶"而是有效的设计辅助工具，并且创造了更多的可能性。6轴机械臂作为人的身体和大脑的拓展，在设计过程中扮演着催化剂作用。

Instead of piecing two distinct workflows together sequentially, the strategy is to explore a method that integrates and unifies 3D-to-2D and 2D-to-3D work. This ambition abandons the paradigm of output-input duality in 2D computer work, instead regarding it as a complete pathway between two physical worlds. This new mode of design breaks down hierarchy and stratification, creating parallelization and integration. It aims to wipe out time-delay, realizing real-time.

The meaning of the robot arms changed from the moment they stepped down from industrial factory assembly lines and were adopted by the SCI-Arc Robot House. We jump out from a

本设计3D到3D的工作流程基于用手势探索人机交互流程。Leap Motion用来读取人的手势动作，然后转译成计算机数据，再由Arduino执行机构将数据输出到机械臂前端的LED装置。人的手势和机械臂的运动路径结合，创造出非预见的运动轨迹。然后，用相机延时摄影，以灯光的形式记录下这种轨迹。

paradigm that regarded them as mechanical devices that obeyed programming and then jogged routinely to repeat a procedure. The robot becomes more than an obedient slave, but a potent auxiliary of the human - a collaborator that cooperates and aids man with design work. It dissolves binary oppositions and dichotomies, instead creating a zone of ambiguity, production and collaboration, visualization and tantalization, mechanization and organization, concretization and abstraction, definition and equivocation, sparking a number of inspirational possibilities. As an extension of the human mind and body, the six axis robot plays the role of a catalyzer in the design process.

The current trajectory of exploration for the 3D-to-3D Workflow is based on gesture, through a prototyping of human motion, as a means of probing the human-robot workflow. Leap motion is used to read hand gestures from the human, which are transferred to the computer as motion data. This data manipulates the Arduino, a control platform, which then outputs information to the electronic LED device, an end effector. The motion from the human gesture is combined with the robot's motion path, creating an un-designed and unpredicted hybrid movement. Thus the process is generative, exceeding its initial movement inputs. As a further step, this hybrid motion is documented visually through the memorization of light paths using a camera. Analysis and future transformations of this light data will point to potential applications of the 3D-to-3D Workflow.

带有手势读取功能的LED装置的机械臂
The LED installation on the robot arm has the function of gesture translation

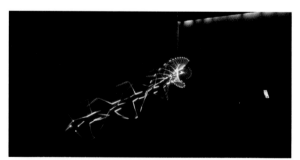
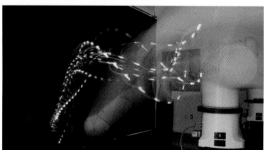
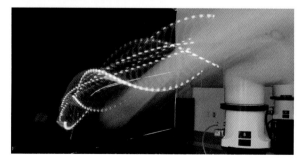

前期试验
Test

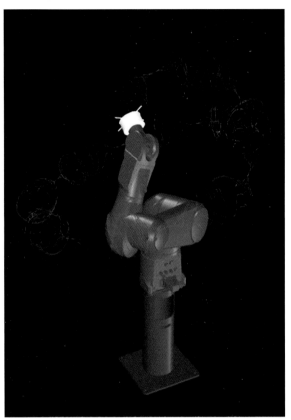
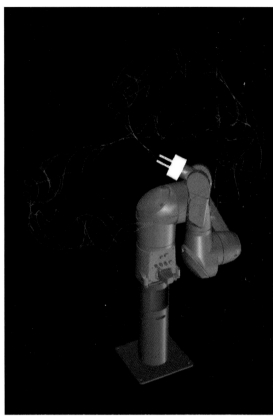

路径模拟
Simulation

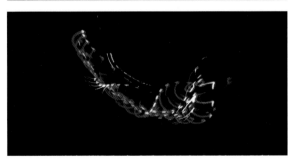
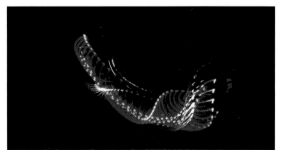

最终成果
Final Model

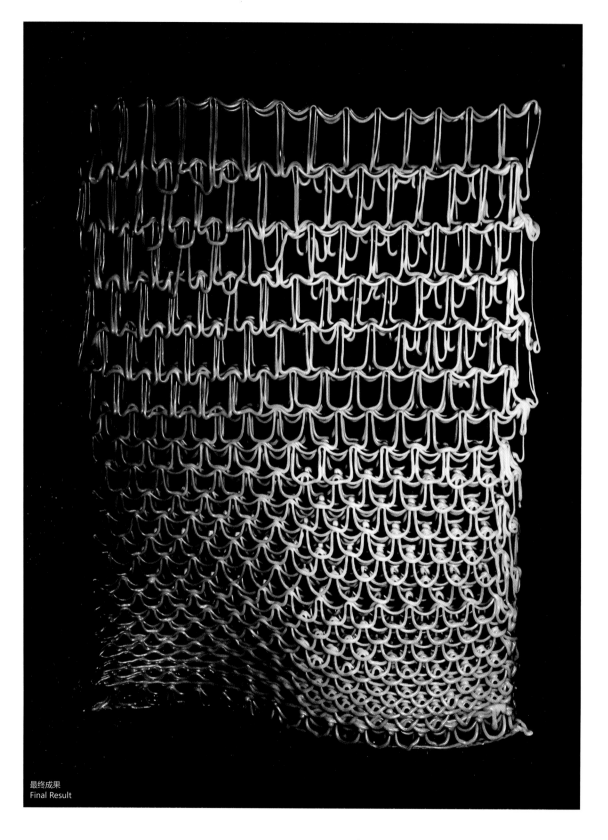

最终成果
Final Result

Domain
Research on Design and Robotic Manufacturing in the Field of Double Curved Geometries

领域
双曲几何领域设计与机器人制造研究

Sandra Manninger, Matias del Campo / Taubman College, University of Michigan
马宁格，马德朴 / 美国密歇根大学Taubman建筑与城规学院

在巴洛克时期，建筑学有时也被理解为应用数学。由伽利略、开普勒、牛顿或是莱布尼茨所带来的自然科学新视野不仅仅被逐渐引入绘画、雕塑中，同时也被广泛运用于塑造当代教堂、宫殿与宅邸的几何图形中。

这个课题开始于一个基于对历史案例和当代理论以及空间概念原理分析的研究项目，逐渐发展为一个设计制造工作室。

第一学期侧重于通过设计进行研究，从历史案例到当代实践，绘制出计算工具在设计与生产领域的应用谱系。在对设计工具与技术进行简短介绍后，这个学期的成果就转变为一个研究项目。双曲率之所以备受关注是由于其在结构、声学、美学以及文化等领域的表现属性。产生出一个系列的设计以此试图理解那些常与建筑品质相关联的属性，伴随着设计评价而来的是规模持续增加的物理模型的制造。

During the Baroque period, architecture was partly understood as applied mathematics. New insights from natural sciences from the likes of Galileo Galilei, Johannes Kepler, Sir Isaac Newton, or Gottfried Wilhelm von Leibniz were introduced not only into paintings and sculptures but also into the geometries that shaped and formed the then contemporary churches, palaces, and mansions.

The thesis started as a research project based on the analysis of historic examples and contemporary theories and principles of spatial concepts to develop into a design and manufacturing studio.

The first semester focused on research through design, drawing a lineage from historic examples to contemporary practices employing computational instruments for design and production.

在实际的建造中,学生需在建筑问题的背景下,借由熟练的技术与工艺创造出一个不使用模具且仅运用单一材料构成的双弧形建筑组成部分。

为深入探索这些必要条件,两个工业机械臂需同步运作。每一个机器人都将被赋予一个提供施工过程二元性的特定任务。其中一个机器人将被装上塑料挤压头,而另一个将被配备有助于纵向建造的电枢。

After a short introduction to design tools and techniques the seminar developed into a research project. Double curvature was scrutinized for its performative qualities in the fields of structure, acoustics, esthetics, and culture. Resulting in a sequence of design studies that tried to understand properties usually associated with architectural qualities. Alongside the design evaluation physical models were manufactured on an ever increasing scale.

In terms of actual fabrication the students were asked to apply the developed skills and techniques to an architectural problem - to create a double curved architectural component constructed from a single material without the use of molds.

To investigate these requisites, two industrial robotic arms were synchronized. Each robot was given a specific task providing duality to the construction process, one robotic arm equipped with a plastic extrusion head, and the other outfitted with an armature to aid vertical construction.

Additive production procedures – such as extruding materials – are unique in the construction environment as they do not subdivide elements but utilize a single gesture that is driven by protocols defined through a relationship between environmental pressures, material properties, the behavior of the tool, and design intention. This robotic gesture can be employed onsite to generate a continuous construction method thus facilitating the building process, and might be considered an alternative to the labor and detail intensive construction methods based on prefabricated components.

圣卡罗教堂的几何分析
Geometric Analysis of San Carlo Alle Quattro Fontane

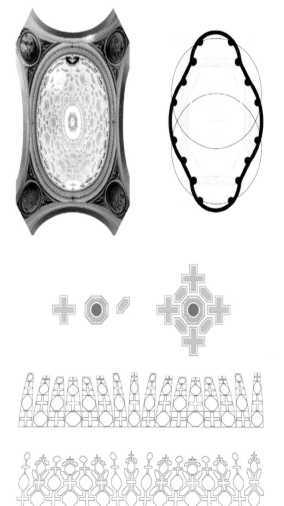

项目团队 Project Team

指导教师:马宁格,马德朴
Instructors: Prof. Sandra Manninger, Prof. Matias del Campo
学生:艾伦•杜夫,乔丹•路特伦,莱恩•斯坎兰
Students: Ellen Duff, Jordan Lutren, Ryan Scanlan

如挤压材料等的附加生产程序在建造环境中十分特殊，因为它们并没有对元素进行细分，而是利用由环境压力关系、材料属性、工具行为以及及设计意图形成的单一的动作。这个机械动作可被现场采用以产生出一个连续性的施工方式，从而促进整个建造过程，甚至可以被视为劳工的替代品以及基于预制组件上的一个细节密集的施工方式。

群体图像和群体意向
Bodies Agglomerations and Feild Condition

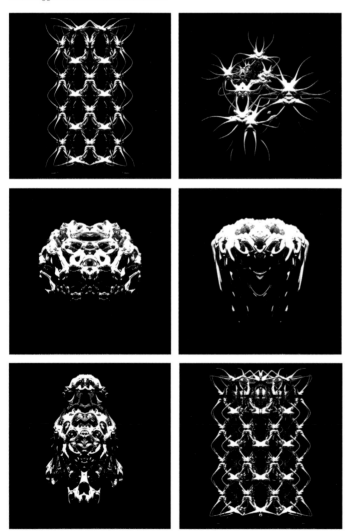

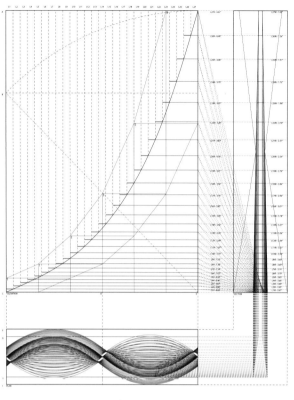

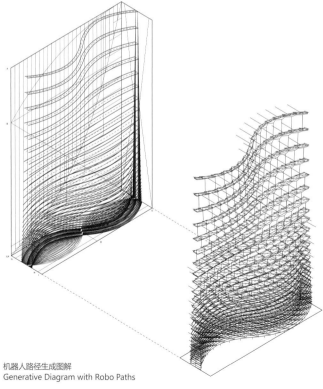

机器人路径生成图解
Generative Diagram with Robo Paths

机器人建造过程
Robotic Fabrication Process

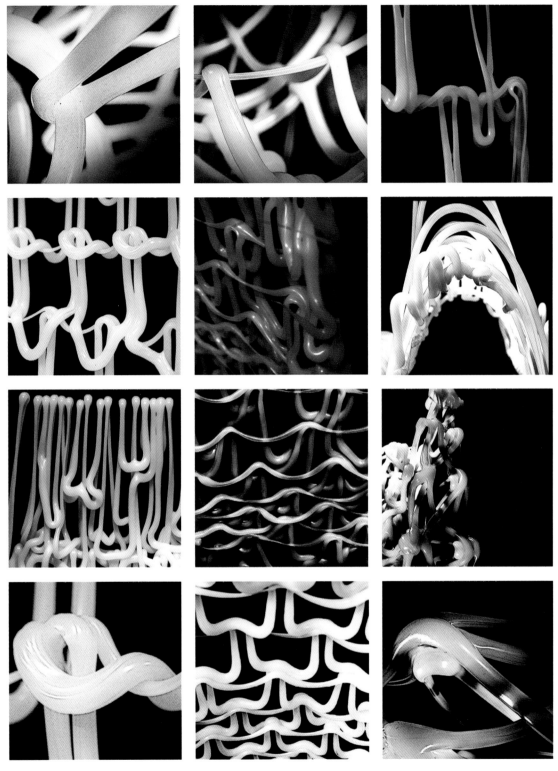

细部
Details

Hypebody研究小组通过优化结构、用机器人3D打印技术建造的1:1尺寸的城市家具片段。
Fragment of urban furniture (1:1 scale) structurally optimized and robotically 3D printed at Hyperbody.

Robotic Building
机器人建筑

Hypebody, TUDelft
荷兰戴尔夫特大学Hypebody研究小组

机器人建筑领域涉及到了物理建造环境以及以机器人为支撑的建造过程的共同支撑。机器人建筑物理环境是一种可重构的适应性的环境氛围,整个响应过程通过能够实现建筑与使用者及周围环境实时互动的传感器 – 促动器机械系统来完成。这个过程要求设计到建造的整个过程(D2P)以及操作链可以实现由机器人来驱动(部分或全部)。随着Hypebody研究小组对于机器人建造概念设计的推进和实践应用,推动了交互式及主动式的建筑构件在不断变化的环境中相互作用的想法实现。这一系列的研究过程都是从生命周期的角度来理解建筑具有的社会经济及生态影响。

机器人建造构件设想的提出或许成为了应对快速增长的人口和城市密集化的可行性解决方案。通过空间重构(在减小的时间帧内实现多重、不断变化的使用)解决建造空间利用率低下(25%)的问题;此外,嵌入式、交互式或机器人能量及气候控制系统在减小建筑生态足迹的同时,能够实现基于时间的、以需求为导向的空间使用过程。此类机器人系统采用将参数化模型与机器人制造及运行装置相连接的D2P运行过程,实现定制构件的高效建造及运行的私人化定制过程。

Robotic building implies both physically built robotic environments and robotically supported building processes. Physically built robotic environments are reconfigurable, adaptive environments incorporating sensor-actuator mechanisms that enable buildings to interact with their users and surroundings in real-time. These require design-to-production (D2P) and operation chains that may be (partially or completely) robotically driven. Hyperbody's design and development of concepts and practical applications for robotic building, leading to the emergence of interactive and proactive building components, which act and interact in ever-changing environments, are based on an understanding of buildings from a life-cycle perspective with respect to their socio-economical and ecological impact.

The assumption is that robotic building components may offer solutions for dealing with the rapid increase of population and urban densification, as well as the contemporary inefficient use (25%) of built space by introducing spatial reconfiguration, which enables multiple, changing uses within reduced timeframes. Furthermore, embedded, interactive or robotic energy and climate control systems may reduce a building's ecological footprint while enabling a time-based, demand-driven use of space. Such robotic systems employ D2P and operation processes that are connecting parametric models with robotic production and operation devices in order to achieve efficient production and operation of custom-made parts for personalized use.

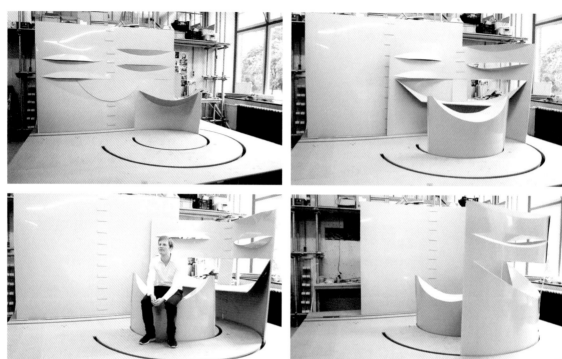

弹出式公寓，2013：Hypebody研究小组与MSc的二年级学生及工业合作伙伴共同开发出的可重构公寓。
Pop-up Apartment, 2013: Reconfigurable apartment developed by Hyperbody with MSc 2 students and industry partners.

Hypebody研究小组在2009年与费斯托合作开发的交互式墙，会对人体运动做出反应。
Interactive Wall developed by Hyperbody in collaboration with Festo responds to people's movement (2009).

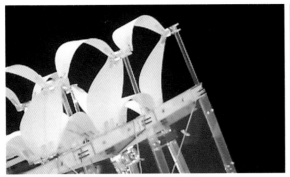
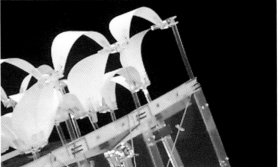

Hypebody研究小组与MSc四年级学生在2012年做的项目,该项目具有通风良好的交互式建筑表皮构件。
Hyperbody MSc 4 project (2012) featuring interactive building skin components employed for ventilation purposes.

在2012年Hypebody研究小组的RDM拱项目中,基于两台大型ABB机器人运行线切割工具,通过机器人建造开发出的原型。
RDM Vault, Hyperbody, 2012: Prototype developed by means of robotic fabrication implemented with two large ABB robots operating wire-cutting tools.

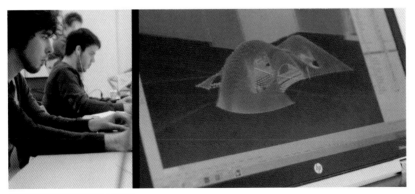

城市家具3D建模与结构模拟(2014)
3D modeling and structural simulation for a piece of urban furniture Hyperbody, TUD (2014)

机器人绘制的分层结构图纸,这种采用坡度、多孔混合材料叠加沉积来打造结构节能建筑,其最初概念由代尔夫特大学Hypebody研究小组提出并进行试验。
Robotic Drawing of layered structure. Gradient, porous, and hybrid materials are architectured and additively deposited for structurally and energy-efficient building (first ideas and tests implemented at Hyperbody, TUD).

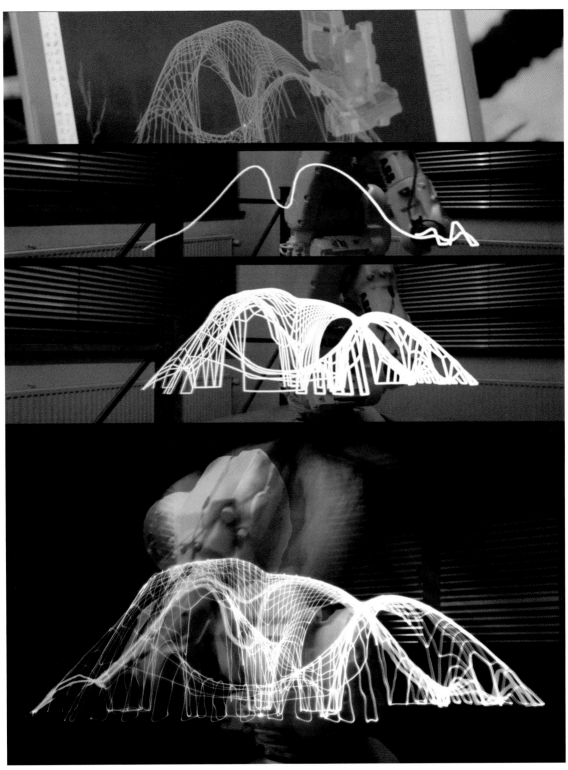

2014年的机器人光试验项目
Robotic Light Experiments, Hypebody, 2014

项目团队

指导教师：徐卫国，于雷，吕帅，刘洁，马逸东
第一组学生：高远，吴晓涵，孙晨炜，王智，马尧，李乐
第二组学生：南天，丁沫，祝豪樱，仇沛然，李同，陈媛
第三组学生：张璐，齐轶昳，彭宁，孙广标，王靖雯，杨鼐

Project Team

Tutors: Xu Weiguo, Yu Lei, Lv Shuai, Liu Jie, Ma Yidong
Group 1 Students: Gao Yuan, Wu Xiaohan, Sun Chenwei, Wang Zhi, Ma Xiao, Li Le
Group 2 Students: Nan Tian, Ding Mo, Zhu Haoying, Chou Peiran, Li Tong, Chen Yuan
Group 3 Students: Zhang Lu, Qi Yidie, Peng Ning, Sun Guangbiao, Wang Qianwen, Yang Ming

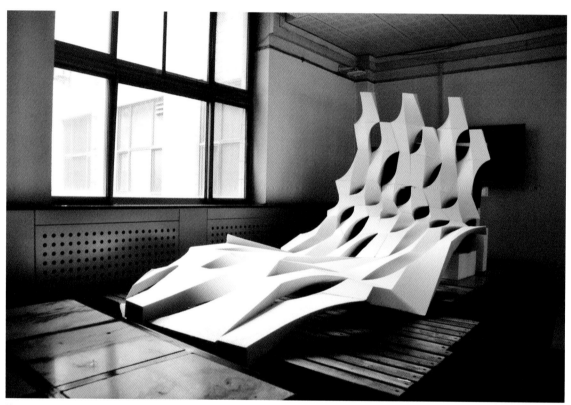

第三组学生作品：植入的韵律
Group 3: Weaving Rhythm

The Exploration on the Teaching of Robotic Fabrication and Design

在机器人建造设计教学上的尝试

Yu Lei, Xu Weiguo / School of Architecture, Tsinghua University
于雷，徐卫国 / 清华大学建筑学院

在徐卫国教授的主导下，清华大学建筑学院2014年秋季学期开设了机器人建造课。这个课程主要是以五年级的设计课为载体，讨论如何将专业课同机器人自主建造技术相结合，将数字设计技术同数字建造技术相结合，并以实体装置为结果的一种教学探索。这种探索在国际上的若干知名设计院校已经逐步展开，但是在国内还是首次以专业设计课的形式进行尝试。

这次设计课的安排一共是15周的时间，将18名本科五年级的学生平均分成三组，每一个组都有相应的研究方向，最后建造三个实体的装置。这18名学生大部分都没有参数化设计的经历，所以基本上都是从零开始。在硬件上，我们采购了一台桌面级的KUKA KR6 900小型6轴机器臂，和一台ABB IRB4600中型6轴机器臂。同学们对于机器臂的原理及操作也基本上是从基础开始学习。

由于我们在软件端采用了Rhinoceros及Grasshopper的参数化工作平台，对于KUKA及ABB等工业级的精密设备都有很好的支持，所以在一定的程度上有效地缩减了设计与硬件之间的磨合期，使这短短的三个月时间达到了很高的效率，并取得了一定的成果。 以下的三组案例，是这次设计课的成果。

In the fall semester of 2014, the School of Architecture in Tsinghua University launched Robotic Fabrication Course under the direction of Professor Xu Weiguo. This course explores the way to combine major design courses with automatic robotic fabrication technology and to integrate digital design with digital fabrication technology. The final result for this is a physical model/installation. This exploration has been made by other international design institutes in the past decade, but this is the first attempt in China to address such issues in combination with a major architectural design course.

The design course lasted 15 weeks, with 18 grade-five students divided equally into 3 groups, and with each group assigned a research direction and the task of building a physical model. Most of the students were without any experience in parametric design, and had to start from scratch. In terms of hardware, we purchased a small desktop-level 6-axis robotic arm KUKA KR6 900 and a medium sized 6-axis robotic arm ABB IRB4600. The students also had to learn the fundamentals and operation knowledge of robotic arms from scratch.

We used Rhinoceros and Grasshopper as the parametric platforms at the software end. They provided strong support for the KUKA and

互动数据与机器建造

这组课题是要求学生将人的身体语言采集下来，随后转换成空间点坐标数据库，并通过一系列算法转换，最终转化为一种艺术化的视觉表现形式。

首先，学生们运用微软的Kinect动作识别器将人在5s内的身体动作通过关节点的空间坐标录制下来，并建立了一个相对完整的空间点阵数据库。接下来，通过对这个数据库的梳理和算法的转换，在计算机内生成空间模型。最后，将空间模型转换成为机械臂的加工路径，并根据加工的要求设计相对应的机器手，然后完成物理模型的最终生成。

由于对数据转换的设计方式不同，这组的学生形成了两个方向：

第一个方向，根据对各个关节点的行为关系的研究，将它们定义为磁力源的南极和北极。这样在人体运动过程中由于各个关节点的空间关系发生了改变，各个磁力源的磁场由于干扰产生了密度和强度的变化。我们在数字模型上可以生动地观察到人体行为的一种全新表现，即运动过程中的一种肉眼不可见的"场效应"现象。这些磁力线的轨迹可以转换为机械臂的运动轨迹代码。最后在机械臂的第6轴顶端安装LED装置，并使用长时间曝光的摄影技术，用光将磁力线的空间轨迹描绘出来。

第二个方向，将一个连续时间间隔的关节坐标点拓扑为一系列二维的剖断面，然后将这些剖面的轮廓线以人的运动轨迹为路

ABB robotic arms, which are sophisticated, industrial standard pieces of equipment. To some extent, this helped to effectively reduce the running period between the coordination of design and hardware. It also improved the quality of the three-month course. The following are the projects of three groups working in this course.

Interactive Data and Robotic Fabrication

The challenge of this group was to collect human body languages in order to coordinate a database of spatial points, and finally transform them to artistic visual presentation via a series of algorithmic translations.

第一组学生作品：互动数据与机器建造
Group 1: Interactive Data and Robotic Fabrication

径进行放样，生成三维模型。将三维模型分成若干块，以适应机器臂的加工范围和运行特点。本方案之所以采用高密度聚苯泡沫作为加工材料，是因为其易于加工、快速成型的特点。通过热切割的方式可以快速地得到我们所需要的模型单元。学生们自己研发了镍铬合金片的高温刮刀，可以在材料表面生成丰富的肌理，通过对切割模式的多次实验，达到了意想不到的表现效果。最后，将模块组合在一起，形成了一个进深接近6m的雕塑。

机器蚕茧

蚕茧的丝状结构貌似是无序的，但是蚕茧的形态却是遵循着自然法则，具备了普遍的相似性。所以说蚕丝的编织方式应该是遵循一种严谨的原则。机器人编织依据算法可具备严谨的规律性，更易于形成可控的表现形态。这组课题的主要概念就是在采用熔融三维打印的技术的支持下，创造出由堆砌"编织的模块"而成的空间结构。

学生们取蚕茧"椭球形"的部分曲面作为具备结构性能的壳体空间结构，将这个壳体结构的表面分格成共面六边形。这里使用了Grasshopper里的Kangaroo模块，对壳体通过使用反重力拓扑算法求得若干个共面六边形。再将其代入优化算法中，将这些六边形重复率提高，从而降低制作过程中六边形模板的数量。

The first step was to complete a spatial lattice database established by transcribing the body movements of one person in the form of spatial coordinates by using a Microsoft Kinect motion capture device. Next, a spatial model was generated through collation of this database and algorithmic translation. Finally, the spatial model was converted to embody a robotic arm processing procedure, and a physical model was thus finally generated through a corresponding robotic hand.

Due to differences in the data conversion, two research approaches were developed.

The first one sought to define articulation points as the magnetic South Pole and North Pole through the study of the behavior of various articulation points. Thus, with any change in the spatial position of each articulation point during human body movement, there is also a change in the density and strength of the magnetic field due to interference. A brand new manifestation of human behavior – such as the "field effect" phenomenon that is invisible to the naked eye during movement – can be vividly observed on a digital model. The pathways of these magnetic lines are translated into a movement path code for the robotic arm, and finally depicted via LED lighting installed on top of the 6-axis robotic arm and captured using long exposure photographic techniques.

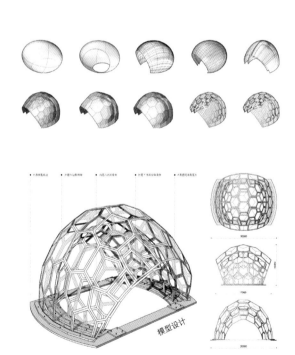

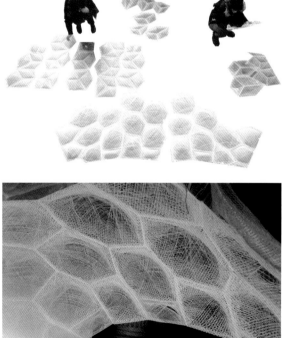

第二组学生作品：机器蚕茧
Group 2: Robotic Silkworm Cocoon

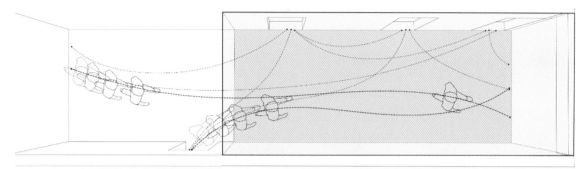

第三组学生作品：植入的韵律，对人流量及其行为特点的概况和分析
Group 3: Weaving Rhythm, summary and analysis on pedestrian flow and characteristics of behaviors

因为三维打印喷头挤出丝的线径只有0.5mm，对于一个以人为尺度的构筑物来说其柔若蚕丝，所以为了提高模块的强度，每个单元体都采用了双层壳体结构，而且内外两层结构采用了嵌入的组合方式，进而提高了结构的整体性。

机械臂在这个不规则多单元的项目中发挥了巨大的作用，学生们只需在Grasshopper中导入每个模板的文件就可自动生成加工路径。经过简单的原点对位工作后，机器人就可独立地完成编织工作。六边形编织单元的路径是经过多次推敲的，目的是为了有效地利用重力的作用使内外层的下垂曲面能够恰好扣合在一起。最后将内层单元体和外层单元体分别缝合成片状后，浮搁在一起，壳体结构可以自支撑地站立，并且富有"天然"的纹理结构。

植入的韵律

这一组的设计主题是基于线切割技术的。机械臂能够灵活准确地控制AB两点连线的空间位置及其角度。如果用加热电阻丝连接AB两点，那么AB两点之间的聚苯泡沫就可以被轻松地切割，并留下光滑的直纹空间曲面。利用这一特性可以掏切出一些复杂的形态。

设计概念源自位于清华大学建筑学院的卫生间、热水房和设计教室之间的植入的一个构筑物，其目的是重新定义这三者之间的空间关系。在对人流量及其行为特点的概括和分析后，通过算法生成了一个具有韵律孔隙的壳状结构，它的三个空间支点安装在这三个入口的门框上，而另一侧则安放于地板上。

这个数字模型被沿着每个开孔的中心点的连线分成若干块，然后求出每个单元块各个面的空间扫描路径，并将路径转换成为机械臂可识别的代码。最后将切割好的单元块积聚成为一个整体。每一个单元体都是从一个比其空间体积略大的长方体中剪切出来的，从一堆碎块中甄别出所需部分的过程充满了乐趣。

The other one sought to topologize the articulation coordinate points at continuous intervals through a series of 2D sections, and to set out contour lines for such sections according to human movement in order to generate a 3D model, which was then divided into several modules in order to meet the processing and operation requirements of the robotic arm. High-density polystyrene foam was selected as the processing material due to its ease of processing and suitability for rapid prototyping. The final model units were fabricated rapidly using hot-wire cutting. Using a high-temperature scraper with nickel-chromium alloy plate developed by students, varied textures were formed on the material surface producing unexpected effects after multiple tests with the cutting mode. Finally, the modules were combined together to form a sculpture with depth of nearly 6m.

Robotic Silkworm Cocoon

The filiform structure of silkworm cocoons might seem somewhat random, but the form of silkworm cocoon follows the laws of nature, and each one is similar to each other. That is to say, the weaving logic of silk worms keep to a kind of rigorous principle. Weaving by robot based on algorithms could be strictly regular and is more likely to produce controllable results. The main concept of this group was to create a spatial structure by aggregating woven modules with the use of fusion three-dimensional printing technology.

The partially curved surface of the ellipsoidal silkworm cocoon was adopted as a shell spatial structure with a certain structural integrity. The surface of this shell structure was then divided into coplanar hexagons. Based on the Kangaroo plug-in for Grasshopper, a number of co-linear hexagons were generated by following the anti-gravitational topological algorithm on the shell. This was then put it into optimization algorithm to improve the repetition rate of these hexagons and reduce the quantity of the hexagon formwork.

As the diameter of the threads extruded from the three-dimensional printing nozzle was only 0.5mm, it seemed like the silk cocoon was comparable to the human scale; therefore, in order to improve the intensity of the module, each unit used a

double-layer shell structure in an embedded combinatory form, thus improving the integrity of the structure.

The robotis arm played a major role in this irregular multi-unit project. Processing routes could be automatically generated only by importing the data of each module into Grasshopper. After simple point matching, the robot could then start weaving independently and automatically. The weaving route of hexagon is calculated to take advantage of gravity in order to make the curved surface of the inner and outer layers to fasten to each other. The units of inner and outer layers are weaved separately weaved to fit together, although the shell structure with its natural texture could stand alone.

Weaving Rhythm

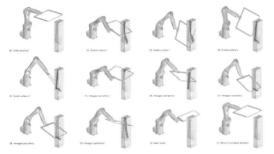

The design topic of this group was based on hot-wire cutting technology. The robotic arms could control the spatial location and degree of the line between point A and B flexibly and accurately. If a resistive heater is used to connect points A and B, the polyphenol foam between point A and B could be easily cut and a smooth straight grain spatial curved surface would be produced. Some complicated forms could also be realized on the basis of this feature.

The design concept is inspired by the insert structure in the space between toilets, the hot water room and classrooms on the third floor of the Architectural School in Tsinghua University. The purpose is to redefine the spatial relationship between them. A shell structure with rhythmic pores is generated using an algorithm based on the summary analysis of pedestrian flows and the characteristics of their behaviors. Three spatial pivot points of the shell structure are installed in the frame of three entrance doors while the last one is placed on floor.

The digital model is divided into several parts along the line connecting the central points of each hole, and the spatial scanning paths of each facet are obtained and transferred into codes that can be identified by the robotic arms. Each unit is cut from a cuboid with a slightly larger spatial volume than the unit and then the cut units are stacked to an integrated form. It is then interesting to select the required units from a stack of components.

第三组学生作品：植入的韵律
Group 3: Weaving Rhythm

基于结构性能的机器人建造
——2014上海"数字未来"暑期工作营

指导老师：于雷，袁烽
学生：史纪，刘浔，罗瑞华，崔宇旗，李晓琳

Robotic Fabrication Based on Structural Performance
—2014 DigitalFUTURE Shanghai Summer Workshop

Tutor: Yu Lei, Philip F. Yuan
Students : Shi Ji, Liu Xun, Luo Ruihua, Cui Yuqi, Li Xiaolin

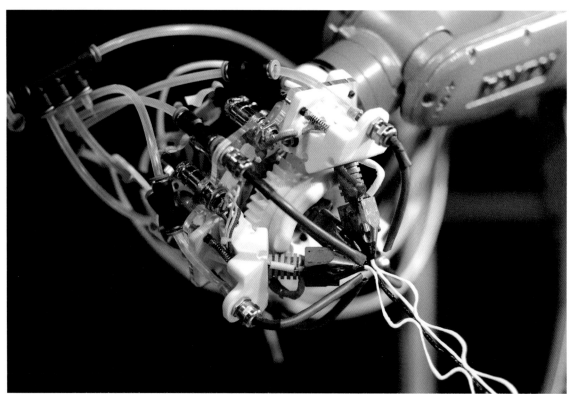

机器人打印工具头
Robotic Printing Tool

Spatial 6D Biomimetic Printing
空间仿生结构6D打印

DigitalFUTURE Shanghai Summer Workshop, Tongji University
同济大学上海"数字未来"暑期工作营

"空间6D打印装置",是一项结合KUKA机器人与3D打印技术的创新。空间打印提供了三维成型的基础,"6D"指的是机器人的6轴,提供了灵活多变的打印路径。传统的3D打印方式依照水平切片打印,而后叠合形成整体模型,这个过程仅仅反映了模型的成型状态,并没有反映结构的自然生长与建构逻辑。

本项目从仿生学的角度出发,学习了蜘蛛吐丝织网的建构过程——一种由一点开始,沿着结构路径生长出复杂空间的自然建构逻辑。我们对传统的打印设备进行了优化:利用KUKA机器臂承载打印工具头,可在空间中沿不同路径打印空间曲线。

与此同时,打印工具头采取了可变的机械设计。4个打印喷头中,3个喷头可以规律性地一张一合,据此可以改变打印过程中线材定型时的截面形状,形成和蜘蛛丝类似、具有结构强度的线材。

这些改动,与其说创新,不如说是回归——当机器臂大幅度运转,打印头一张一合、空间结构沿着三维路径慢慢生长时,正是机器技术向自然建构的回归。

Spatial 6D Biomimetic Printing is an innovation combining the technology of KUKA robot and 3D Printing. 3D printing provides the possibility for instant modeling, and 6D refers to the 6 axes of the robots, which make the way of printing more flexible. There are great limitations with traditional 3D printing methods, which realize a model by making horizontal layers and piling up materials. This method might easily reveal the result of the model, but cannot capture the process and the logic of the natural growing of the structure.

So, our project uses biomimetic approaches as practical strategies to simulate a spider spinning its web, which starts with a point in the space, and grows into a natural, complex and self-supporting structural system. To realize our idea, we improved the traditional 3D printer by applying it to KUKA robot. With the added arm, the 3D printer can easily print any spatial curves without any limitation. Also, the 3D printer itself is improved in order to strengthen the stability of the curve. We combine 4 printers together, 3 of which can switch on and off regularly. As a result the section of the printed curve can be changed, forming a biomimetic string as strong as a spider's thread.

These improvements aim at returning to the original - to nature itself. As the arms of KUKA robot operate, as the printers move, and as the spatial structure begins to sprawl along the track, the high-tech machines return to nature.

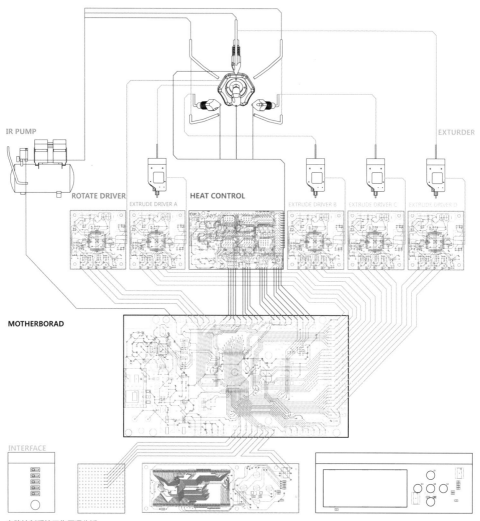

电路控制系统工作原理分析
Control System Working Mechanism

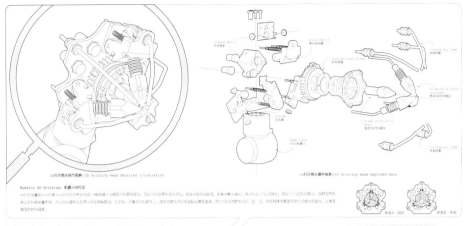

3D打印喷头细节图解 | 3D Printing Head Detailed Illustration   3D打印喷头爆炸轴测 | 3D Printing Head Exploded Axon

Robotic 6D Printing: 机器人6D打印

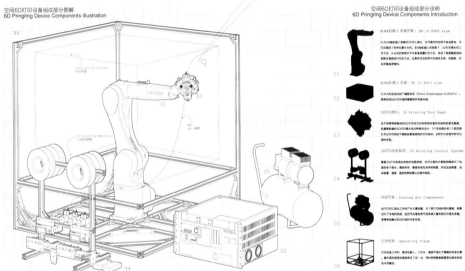

空间6D打印设备组成部分图解
6D Pringitng Device Components Illustration

空间6D打印设备组成部分说明
6D Pringitng Device Components Introduction

机械工作系统工作原理分析
Working System Working Mechanism

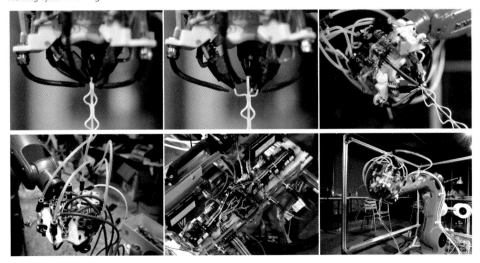

201

旮旯酒吧建筑师计划座椅设计

设计师：袁烽
设计团队：闫超，孟浩

Chair design for GALA Bar

Designer: Philip F. Yuan
Design Team: Yan Chao, Hyde Meng

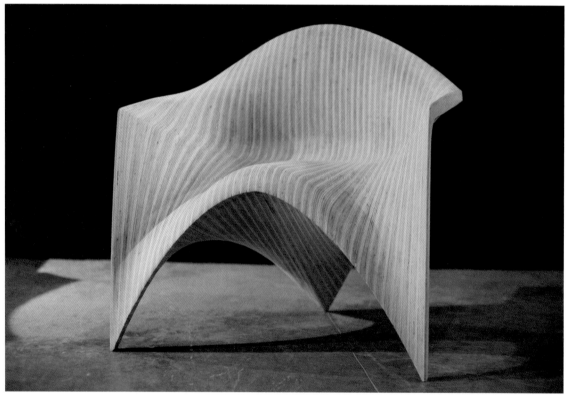

边锋椅的设计以曲面边缘作为几何操作的核心
In the design process, edge manipulation as the main subject integrates the elements of the chair

Edge Chair
边锋椅

Digital Design Research Center (DDRC), CAUP, Tongji University
同济大学建筑与城市规划学院数字设计研究中心

边锋椅，作为使用者与建筑之间的次级媒介，将建筑几何的定义向人体尺度进行拓展，在形式性能和物质建构两个方面探讨了人体与几何、机器生产和个性定制之间的互动关系。在几何性能化方面，设计以曲面边缘作为几何操作的核心，将座椅的支撑、坐垫、扶手和靠背整合在连续的光滑曲面之中，通过座椅边缘对人体坐姿的回应，建立起人体工程学与几何形式之间的联系。在建构过程中，机器人作为加工媒介对木料进行等高线边缘切割，将木材的"柔软"特性最大化，实现了人体触感与木材表面完美贴合，同时消除了个体偏好与批量生产之间的矛盾。

Edge Chair, as a media between human and architecture, is trying to readdress the meaning of geometry in human scale and to redefine the relation between machine production and customization in both form generation and material fabrication aspects. In the design process, edge manipulation as the main subject integrates the elements of the chair, such as supports, cushion, and hand rail into a continuous surface. And by following the curvature of human body, the connection between human engineering and performative geometry are established. In the fabrication aspects, robot as a digital-controlled platform maximizes the "softy" of wood through a series of edge cutting process, which resolves the conflict between personal preference and batch production finally.

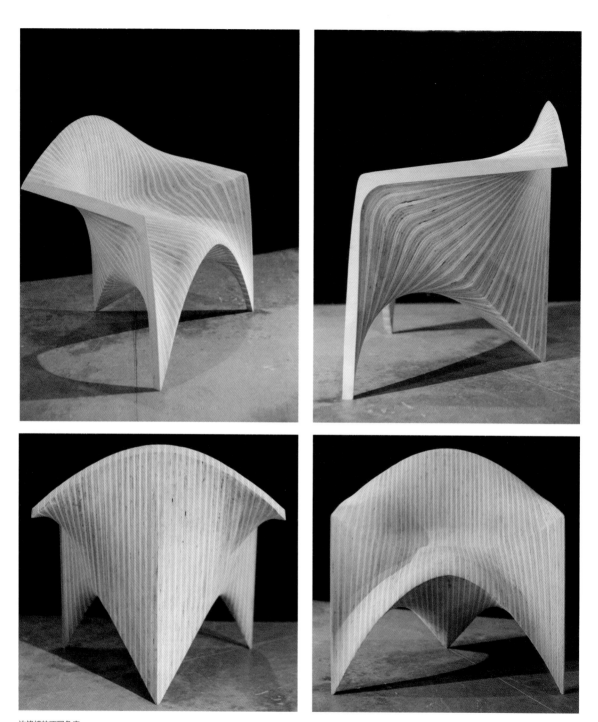

边锋椅的不同角度
Different Views of the Chair

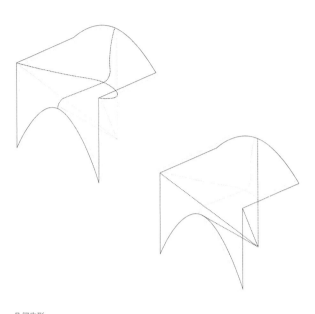

几何生形
Geomotrical Formation

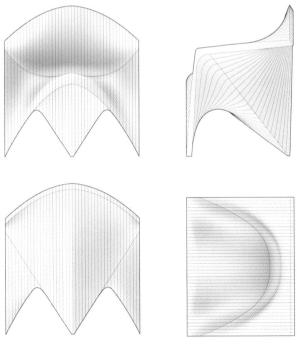

数字模型
Digital Model

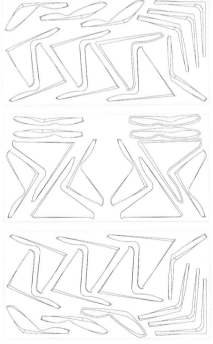

加工模块
Fabrication Components

2015年，拓扑表皮在第十届上海双年展城市馆"人机未来"展上展出
Topological Surface was exhibited in the "Robotic Future" exhibition of 10th Shanghai Biennale, 2015

Topological Surface
拓扑表皮

Digital Design Research Center (DDRC), CAUP, Tongji University
同济大学建筑与城市规划学院数字设计研究中心

"拓扑表皮"项目尝试以机器人作为材料加工媒介，对数控建造环境下复杂形式建构的可能性进行探索。项目研究以Seifert曲面——一种具有方向性且边界盘绕的复杂曲面为设计原型，通过对三条曲面边缘的套嵌关系进行操作，创造出超越第三维度限制的复杂拓扑空间形式。在建造过程中，拓扑曲面被细分成固定尺寸的较小单元，之后依托数控机器的精确性，对每个单元体进行纤维混凝土的非在场浇筑，并最终通过一套混凝土预制装配体系，在现场组合成传统建造方式难以实现的复杂形态。

Topological Surface tries to explore the possibility of using robots as the material processing instrument to handle complicated structures under the environment of numerical control architecture. With Seifert surface (whose boundary is a given knot or link) as its design prototype, the project deals with interlocking relations of three curving surfaces, and creates a complicated topological form that breaks restrictions of the three-dimensional space. During the building process, curving surfaces of the topology were divided into smaller units of fixed sizes. The units were created with fiber cement outside the exhibition venues, using the accuracy of Numerical Control Machines. They were assembled on-site through a solution designed for precast concrete modules, to achieve the complexity impossible for traditional building patterns.

拓扑表皮
Topological Surface

设计：袁烽，闫超
作品尺寸：8.4m×5.6m×4.7m
建造支持：上海斯诺博金属构件发展有限公司

Design: Philip F. Yuan, Yan Chao
Size: 8.4m × 5.6m × 4.7m
Fabrication Support: Sinobau

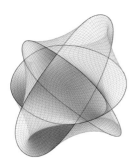

几何生形
Geometrical Formation

加工模块切分
Fabrication Components

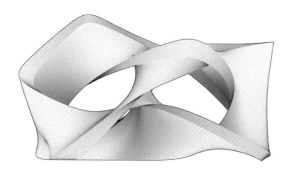

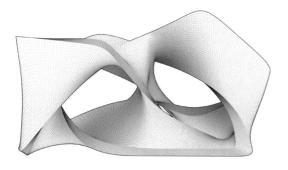

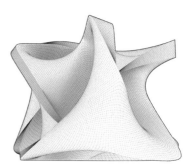

立面图
Elevations

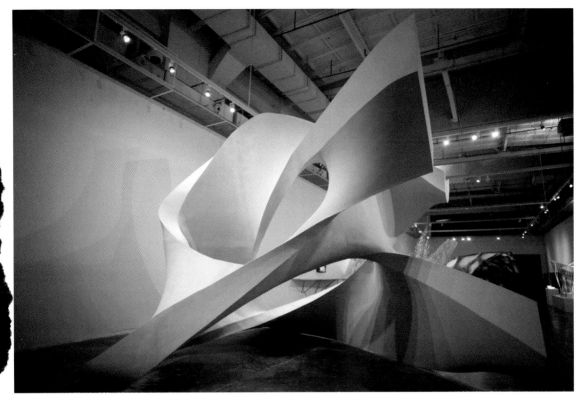

2015年，拓扑表皮在第十届上海双年展城市馆"人机未来"展上展出
Topological Surface was exhibited in the "Robotic Future" exhibition of 10th Shanghai Biennale, 2015

BIOGRAPHIES
作者简介

袁烽 / Philip F. Yuan

袁烽，2003年获得同济大学建筑城规学院建筑设计与理论博士学位，在2008~2009年期间，作为访问学者到美国麻省理工大学进行了交流学习。近年来主要致力于数字设计方法与数字化建造方法的研究与实践，是国家教育部高密度人居环境实验室"数字设计研究中心"的负责人，也是"中国建筑学会建筑师分会——数字建筑设计专业委员会"的发起人之一。

袁烽多次荣获国内外设计领域的重要奖项，其中包括：2015年WA建筑成就奖；2014年中国建筑学会"青年建筑师"奖、2014年亚洲建协的年度建筑优秀入围奖；2014年欧洲维纳博艮砖筑奖；2013年同济大学冯纪忠建筑教育奖等。近年来他也参加并组织了多次国内外重要建筑学术展览，其中包括：2014年，作为学术主持，策划了第十届上海双年展城市馆的"人机未来"展览；2013年，深港双年展；2013年，米兰建筑三年展的"从研究到设计——同济建筑师展"；2013年，西岸建筑与当代艺术双年展；2012年，欧洲太阳能十项全能竞赛（Solar Decathlon Europe 2012）并获得"环境性能测试"以及"功能布局"两项大会三等奖。2011年以来，他作为每年由同济大学建筑城规学院举办的上海"数字未来"系列学术活动的组织者之一，成功促进了同济与全球各大院校在数字设计领域的交流合作。2013年袁烽受SmartGeometry邀请，参加在香港大学举行的国际学术会议，并发表了题为"Performative Geometry"的主题演讲；2014年受邀参加在密西根大学举办的"Robotic Fabrication in Architecture，Art and Design2014"国际会议并做题目为"Performative Tectonics"的大会主题演讲。2013年7月，袁烽在AU Space举办了"自主建构——研究·教学·实践"的建筑个展，这是他在数字设计展览方向的又一个探索。

袁烽已出版的学术著作有《现实建构》（中国建筑工业出版社，2011年出版）、《观演建筑设计》（同济大学出版社，2012年出版），与尼尔·里奇共同编著中英双语图书《建筑数字化建造》和《建筑数字化编程》（同济大学出版社，2012年出版），以及《探访中国数字建筑设计工作营》（同济大学出版社，2013年出版），曾担任《时代建筑》"今日建筑"专栏主持、客座编辑，并在《建筑学报》、《新建筑》和《城市建筑》等多本专业期刊上发表大量学术论文。

Philip F. Yuan is an academic researcher and architect, who focus on digital design and fabrication methodology. He received his PhD from Tongji University in 2003 and had been a visiting scholar in MIT during 2008~2009. In recent years, Yuan co-founded the Digital Architectural Design Association (DADA) of The Architectural Society of China (ASC), and has been working as the coordinator of Digital Design Research Center (DDRC) in the College of Architecture and Urban Planning(CAUP) at Tongji University.

As a pioneer in architecture, he has won many international awards, such as *2015 WA Award*, *2014 Arcasia Award*, *2014 Wienerberger Brick Award*, *2014 ASC's Young Architect Award* and *2013 Tongji's Feng Jizhong Architectural Education Award*. He was invited to give keynote speeches and attend many international and domestic academic conferences and exhibitions. He was the academic host of "*Robotic Future*" Exhibition in 10th Shanghai Biennale City Pavilion Section. The exhibition he has been invited to participate includes "*From Research to Design-Tongji Architects*" Exhibition at Milan Triennial 2012, "*Solar Decathlon Europe 2012*" in Madrid and won the third prizes of "*Best Environment Performance*" and "*Best Function Layout*". He was invited to giving a keynote speech about "*Performative Geometry*" at SmartGeometry 2013 Conference in Hong Kong University and also invited to give a speech about "*Perormative Tectonics*" at "*Robotic Fabrication in Architecture: Art and Design 2014*" Conference in University of Michigan. As the organizer of DigitalFUTURE Shanghai joint summer school program, Yuan has been strongly promoting digital design and fabrication since 2011. In July 2013, his solo exhibition "*AUTONOMOUS TECTONICS - Research·Teaching·Practice*" was hold in AU Space, it shows his exploration on digital design exhibition field.

He is the author of *A Tectonic Reality* (China Architecture & Building Press, 2011), *Theater Design* (Tongji University Press, 2012), co-editor (with Neil Leach) of *Fabricating the Future* and *Scripting the Future* (Tongji University Press, 2012), co-editor (with Neil Leach) of *Digital Workshop in China*. He used to work as the guest editor of *T+A Magazine*. He has published many articles on *Architectural Journal*, *New Architecture* and *U+A Magazine*.

阿希姆·门格斯 / Achim Menges

阿希姆·门格斯，注册建筑师，作为德国斯图加特大学建筑系教授，他于2008年创建了斯图加特大学计算设计学院。2009年起，他在哈佛大学设计学院担任客座教授。门格斯教授于2002年作为优秀毕业生从伦敦建筑联盟学校（AA）毕业，2002~2009年他作为指导教师在AA的"涌现技术与设计"专业授课，2009~2012年作为该专业的客座教授任职。

门格斯教授的研究领域主要在计算设计的整合设计过程的发展上，包括仿生工程、参数化设计、进化算法和计算机辅助建造的交叉研究。他的作品都是建立在与结构工程师、计算机专家、材料专家和生物学家的跨学科的合作基础上。阿希姆·门格斯在其相关设计领域出版了数本著作，发表了百余篇学术论文。

他的作品多次荣获了国际奖项，被国际媒体报道，并多次参与了国际展览，他的作品也被包括巴黎蓬皮杜中心在内的许多世界知名博物馆收藏。

Achim Menges is a registered architect and professor at the University of Stuttgart, where he is the founding director of the Institute for Computational Design at the University of Stuttgart since 2008. In addition, he is Visiting Professor in Architecture at Harvard University's Graduate School of Design since 2009. He graduated with honours for the best diploma of 2002 from the AA School of Architecture in London, where he subsequently taught as Studio Master of the Emergent Technologies and Design Graduate Program from 2002 to 2009, as visiting professor from 2009 to 2012 and as Unit Master of Diploma Unit 4 from 2003 to 2006.

Menges' practice and research focus on the development of integral design processes at the intersection of design computation, biomimetic engineering and robotic manufacturing that enables a performative and sustainable built environment. His work is based on an interdisciplinary approach in collaboration with structural engineers, computer scientists, material scientists and biologists. He has published several books on this work and related fields of design research. In addition he is the author/coauthor of more than 100 scientific papers and numerous articles.

His projects and design research has received many international awards, has been published and exhibited worldwide, and form parts of several renowned museum collections, among others, the permanent collection of the Centre Pompidou in Paris.

尼尔·里奇 / Neil Leach

尼尔·里奇是欧洲研究生学院的教授，哈佛大学设计学院和同济大学建筑与城市规划学院的客座教授。同时他也是南加州大学的兼职教授和美国宇航局（NASA）先进概念项目的成员。里奇教授近期正在开展由NASA资助的一个研究项目，该项目主要是探索在月球和火星上运用机器人建造技术打印建筑结构的可能。他从事建筑理论研究，已出版了25本书籍，这些书籍被译成6种语言在全世界发行。他在建筑理论上的著作包括《反思建筑》（Routledge出版社，1997年出版）、《建筑美学》（MIT出版社，1999年出版）、《千年文化》（Elipsis出版社，1999年出版）、《空间的象形文字》（Routledge出版社，2002年出版）、《遗忘海德格尔》（Paideia出版社，2006年出版）、《伪装》（MIT出版社，2006年出版）等。他在数字设计方面编著的出版物包括《数字世界》（Wiley出版社，2002年出版）、《数字建构》（Wiley出版社，2004年出版）、《数字现实：学生建筑设计作品》（中国建筑工业出版社，2010年出版）、《建筑数字化建造》（同济大学出版社，2012年出版）、《建筑数字化编程》（同济大学出版社，2012年出版）、《设计智能：高级计算性建筑生形研究学生建筑设计作品》（中国建筑工业出版社，2013年出版）、《集群智能》（同济大学出版社，即将出版）等。

Neil Leach is Professor at the European Graduate School, Visiting Professor at Harvard GSD and Tongji University, Adjunct Professor at USC, and a NASA Innovative Advanced Concepts Fellow. He is currently working on a research project funded by NASA to develop a robotic fabrication technology to print structures on the Moon and Mars. He has published 25 books, which have been translated into 6 other languages. His publications on architectural theory include *Rethinking Architecture* (Routledge, 1997), *The Anaesthetics of Architecture* (MIT Press, 1999), *Millennium Culture* (Ellipsis, 1999), *Architecture and Revolution* (Routledge, 1999), *The Hieroglyphics of Space* (Routledge, 2002), *Forget Heidegger* (Paideia, 2006) and *Camouflage* (MIT Press, 2006). His publications on computational design include *Designing for a Digital World* (Wiley, 2002), *Digital Tectonics* (Wiley, 2004), *Machinic Processes* (CABP, 2010), *Fabricating the Future* (Tongji UP, 2012), *Scripting the Future* (Tongji UP, 2012), *Design Intelligence: Advanced Computational Research* (CABP, 2013) and *Swarm Intelligence* (Tongji UP, forthcoming).

图书在版编目（CIP）数据

建筑机器人建造：汉文、英文 /
袁烽, (德) 门格斯 (Menges,A.) , (英) 里奇 (Leach,N.) 等著.
-- 上海：同济大学出版社, 2015.6（数字设计前沿系列丛书 / 袁烽, 江岱主编）
ISBN 978-7-5608-5845-6

Ⅰ.①建... Ⅱ.①袁... ②门... ③里... Ⅲ.①建筑机器人 - 制作 - 汉、英 Ⅳ.①TP242.3

中国版本图书馆CIP数据核字(2015)第108181号

建筑机器人建造
Robotic Futures

袁　烽　（德）阿希姆·门格斯　（英）尼尔·里奇　等著
Philip F. Yuan　Achim Menges　Neil Leach

责任编辑　江　岱
助理编辑　武　蔚
责任校对　徐春莲
装帧设计　袁佳麟，4aTEAM (www.4ateam.com)

出版发行　同济大学出版社
　　　　　（www.tongjipress.com.cn　地址：上海市四平路1239号　邮编：200092　电话：021-65982473）
经　　销　全国各地新华书店
印　　刷　上海盛隆印务有限公司
开　　本　787mm×1092mm　1/16
印　　张　13.25
印　　数　1-2100
字　　数　331000
版　　次　2015年6月第1版　2015年6月第1次印刷
书　　号　ISBN 978-7-5608-5845-6
定　　价　100.00元

本书若有印刷质量问题，请向本社发行部调换
版权所有　侵权必究